WITHDRAWN

D1044912

Current Topics in Microbiology

Microbiology

196 and Immunology

Editors

A. Capron, Lille · R.W. Compans, Atlanta/Georgia
M. Cooper, Birmingham/Alabama · H. Koprowski,
Philadelphia · I. McConnell, Edinburgh · F. Melchers, Basel
M. Oldstone, La Jolla/California · S. Olsnes, Oslo
M. Potter, Bethesda/Maryland · H. Saedler, Cologne
P.K. Vogt, La Jolla/California · H. Wagner, Munich
I. Wilson, La Jolla/California

Current Topics in
Microbiology
and Immunology 196

The Role of Nitric Oxide in Physiology and Pathophysiology

Edited by H. Koprowski and H. Maeda

With 21 Figures and 7 Tables

Springer-Verlag

Berlin Heidelberg New York
London Paris Tokyo
Hong Kong Barcelona
Budapest

Hilary Koprowski
Department of Microbiology and Immunology
Head of the Center of Neurovirology
at Thomas Jefferson University
1020 Locust Street
Philadelphia, PA 10107
USA

Hiroshi Maeda
Department of Microbiology
Kumamoto University School of Medicine
Honjo 2-2-1
Kumamoto 860
Japan

Cover Illustration: This drawing illustrates the multi-faceted function of Nitric Oxide in the human organism and lists, at the same time, known inhibitors of Nitric Oxide synthase and a scavenger substance (PTIO) which binds directly to NO. Picture by courtesy of Neilson Carlin.

Cover design: Künkel und Lopka, Ilvesheim

ISSN 0070-217X
ISBN 3-540-58214-2 Springer-Verlag Berlin Heidelberg New York
ISBN 0-387-58214-2 Springer-Verlag New York Berlin Heidelberg

This work is subject to copyright. All rights are reserved, whether the whole or part of the material is concerned, specifically the rights of translation, reprinting, reuse of illustrations, recitation, broadcasting, reproduction on microfilm or in any other way, and storage in data banks. Duplication of this publication or parts thereof is permitted only under the provisions of the German Copyright Law of September 9, 1965, in its current version, and permission for use must always be obtained from Springer-Verlag. Violations are liable for prosecution under the German Copyright Law.

© Springer-Verlag Berlin Heidelberg 1995
Printed in Germany

The use of general descriptive names, registered names, trademarks, etc. in this publication does not imply, even in the absence of a specific statement, that such names are exempt from the relevant protective laws and regulations and therefore free for general use.

Product liability: The publishers cannot guarantee the accuracy of any information about dosage and application contained in this book. In every individual case the user must check such information by consulting other relevant literature.

Production: PRODUserv Springer Produktions-Gesellschaft, Berlin
Typesetting: Thomson Press (India) Ltd, New Delhi
SPIN: 101 28 452   27/3020-5 4 3 2 1 0 – Printed on acid-free paper.

Preface

Fascination with nitric oxide dates back only a few years to when scientists discovered that this, the smallest known molecule with biological functions, has many properties. A wide range of activities attributed to nitric oxide, which is a product of the two enzyme constituent nitric oxide synthase and induced nitric oxide synthase, have been intensely studied by scientists throughout the world.

The present volume contains papers presented at a workshop on "The Role of Nitric Oxide in Physiology and Pathophysiology" at Thomas Jefferson University, Philadelphia, Pennsylvania, on October 25, 1993. The authors came from different parts of the world to participate in the workshop and presented up-to-date results of their research into nitric oxide. The editors would like to express their thanks for these contributions which will enrich the literature on the importance of nitric oxide in physiology and pathology.

HILARY KOPROWSKI
HIROSHI MAEDA

List of Contents

List of Contributors

(Their addresses can be found at the beginning of their respective chapters.)

Inducible Nitric Oxide Synthase: Regulation Subserves Function

C. NATHAN

1 Inducible Isoform of Nitric Oxide Synthase Is a Distinct Gene Product

Nitric oxide synthase (iNOS) was purified (STUEHR et al. 1991) and cloned (XIE et al. 1992) from interferon-γ (IFNγ)- and lipopolysaccharide (LPS)-activated mouse macrophages. The primary structure, deduced from the cDNA and confirmed by HPLC-electrospray ionisation mass spectrometry, established that this inducible enzyme differs from the constitutive NOSs (cNOSs) cloned from cerebellar and endothelial libraries. Western blotting detected a highly similar enzyme in a wide variety of cell types after activation, including cardiac myocytes, where NO appeared to suppress rhythmicity (ROBERTS et al. 1992). Southern blotting and genomic cloning (CHARTRAIN et al.1994) strongly suggest that there is a single iNOS gene. Thus, iNOS is a unique gene that can be expressed widely following activation.

Cornell University Medical College, Box 57, 1300 York Ave, New York, NY 10021, USA

2 Inducible Nitric Oxide Synthase Binds Calmodulin Without Dependence on Elevation of Calcium

Catalysis by cNOSs depends on binding of calmodulin (CaM), which in turn depends on an elevation of Ca^{2+} ($EC_{50} \geq 200$ nM) above the levels typically found in resting cells (≤ 100 nM). In contrast, pure iNOS functions without addition of exogenous CaM or Ca^{2+}, despite containing a canonical CaM-binding motif (amino acid residues 503–532). This paradox was explained by the ability of iNOS to bind CaM as a subunit at trace levels of Ca^{2+} (39 nM). CaM was identified in pure iNOS by reactivity with anti-CaM mAb, migration on SDS-PAGE, reverse phase-HPLC retention time, tryptic map, partial amino acid sequence, and electrospray ionization mass spectrometry (Cho et al. 1992).

3 Synergistic Transcriptional Induction of Inducible Nitric Oxide Synthase Is Mediated Through an Upstream Regulatory Region Rich in Transcription Factor Binding Site Consensus Sequences: Key Role of Nuclear Factor-κB

Instead of being regulated by agonist-induced Ca^{2+} transients, iNOS is controlled chiefly by transcription, as shown by the dependence of iNOS enzyme activity, iNOS mRNA synthesis, and iNOS antigen expression on exposure of macrophages to a synergistic combination of signals, such as IFNγ and LPS, and the sensitivity of each of these processes to nontoxic concentrations of actinomycin D. Induction of iNOS in mouse peritoneal macrophages by IFNγ and LPS required synthesis of an intermediary protein(s) and may involve tyrosine phosphorylation (Xie et al. 1993). From a mouse genomic library, we cloned a 1749 base pair fragment from the 5′ flanking region of the iNOS gene and used S1 nuclease mapping and primer extension to identify the mRNA transcription start site within it. The mRNA initiation site is preceded by a TATA box and at least 22 oligonucleotide elements homologous to consensus sequences for the binding of transcription factors involved in the inducibility of other genes by cytokines or bacterial products. These include ten copies of IFNγ response element (γ-IRE); three copies of γ-activated site (GAS); two copies each of nuclear factor-κB (NF-κB), IFN-α-stimulated response element (ISRE), activator protein-1 (AP-1), and tumor necrosis factor response element (TNFRE); and one X box. Plasmids in which all or the downstream one-half or one-third of this region of iNOS were linked to a reporter gene encoding chloramphenicol acetyltransferase were transfected into RAW264.7 macrophage-like cells. All these constructs conferred inducibility of the iNOS promoter by LPS, but only the construct containing all

1749 base pairs conferred synergistic inducibility by IFNγ plus LPS (XIE et al. 1993). Analysis of further constructs indicates that the downstream NF-κB site in the iNOS promoter, NF-kBd, is necessary to confer inducibility by LPS. A nuclear protein which binds to NF-kBd after LPS-induction was identified as an NF-kB heterodimer by competition, supershift, and UV cross-linking, and induction of iNOS mRNA was blocked by the NF-kB inhibitor, pyrollidine dithiocarbamate (XIE et al. 1994).

4 Antiviral Activity of Inducible Nitric Oxide Synthase: Potential Explanation for Widespread Inducibility of an Autotoxic Enzyme

It is a mystery why evolution has conferred on so many types of cells the capacity to respond to immune or inflammatory signals with the induction of an enzyme that, once tnanslated, can function for days to generate large amounts of a potentially toxic radical. We reasoned that NO may defend against a class of pathogens by which many cells types can be infected and may be a toxin against which it is difficult for pathogens to evolve to a resistant state. Consistent with this view, KARUPIAH et al. (1993) demonstrated that iNOS can account for a major portion of the antiviral activity of IFN-γ in mouse macrophages infected with ectromelia, herpes simplex or vaccinia viruses and is critical to the recovery of mice from ectromelia. The mechanism has not been defined, but one key target molecule may be ribonucleotide reductase (KWON et al. 1991).

5 Inducible Nitric Oxide Synthase Is Also Suppressible

The potential autotoxicity of iNOS for the host is exemplified by its critical role in experimental arthritis (MCCARTNEY-FRANCIS et al. 1993). Thus, it is not surprising that the widespread capacity to induce iNOS is balanced by an equally widespread ability of host cells to produce cytokines that can antagonize its induction. These include transforming growth factor-β (TGF-β), interleukin-4, and macrophage deactivation factor (DING et al. 1990; VODOVOTZ et al. 1993; BOGDAN et al. 1994). The potency of TGF-β in this regard may result from the multiplicity of levels at which it suppresses iNOS, including decreased stability and translation of iNOS mRNA and decreased stability of iNOS protein (VODOVOTZ et al. 1993).

In summary, the fact that iNOS is regulatable at transcriptional and multiple post transcriptional levels may reflect the important dual role of this enzyme in host defense and autotoxicity.

References

Bogdan C, Vodovotz Y, Paik J, Xie QW, Nathan C (1994) Mechanism of suppression of nitric oxide synthase expression by interleukin-4 in primary mouse macrophages. J Leuk Biol 55: 227–233

Chartrain NA, Geller DA, Koty PP, Sitrin NF, Nussler AK, Hoffman EP, Billiar TR, Hutchinson NI, Mudgett JS (1994) Molecular cloning, structure and chromsomal localization of the human inducible nitric oxide synthase gene. J Biol Chem 269: 6765–6772

Cho HJ, Xie QW, Calaycay J, Mumford RA, Swiderek KM, Lee TD, Nathan C (1992) Calmodulin is a subunit of nitric oxide synthase from macrophages. J Exp Med 176: 599–604

Ding A, Nathan CF, Graycar J, Derynck R, Stuehr DJ, Srimal S (1990) Macrophage deactivation factor and transforming groth factors-β-1, -2, and -3 inhibit induction of macrophage nitrogen oxide synthesis by interferon-γ. J. Immunol 145: 940–944

Karupiah G, Xie QW, Buller RML, Nathan C, Duarte C, MacMicking J (1993) Inhibition of viral replication by interferon-γ-induced nitric oxide synthase. Science 261: 1445–1448

Kwon NS, Stuehr J, Nathan CF (1991) Inhibition of tumor cell ribonucleotide reductase by macrophage-derived nitric oxide. J Exp Med 174: 761–768

McCartney-Francis N, Allen JB, Mizel DE, Albina JE, Xie QW, Nathan CF, Wahl SM (1993) Suppression of arthritis by an inhibitor of nitric oxide synthase. J Exp Med 178: 749–754

Roberts AB, Vodovotz Y, Roche NS, Sporn MB, Nathan CF (1992) Role of nitric oxide in antagonistic effects of transforming growth factor-β and interleukin-1β on the beating rate of cultured cardiac myocytes. Mol Endocrinol 6: 1921–1930

Stuehr DJ, Cho HJ, Kwon NS, Weise M, Nathan CF (1991) Purification and characterization of the cytokine-induced macrophage nitric oxide synthase: an FAD- and FMN- containing flavoprotein. Proc Natl Acad Sci USA 88: 7773–7777

Vodovotz Y, Bogdan C, Paik J, Xie QW, Nathan C (1993) Mechanisms of suppression of macrophage nitric oxide release by transforming growth factor-β. J Exp Med 178: 605–164

Xie QW, Cho H, Calacay J, Mumford RA, Swiderek KM, Lee TD, Ding A, Troso T, Nathan C (1992) Cloning and characterization of inducible nitric oxide synthase from mouse macrophages. Science 256: 225–228

Xie QW, Whisnant R, Nathan C (1993) Promoter of the mouse gene encoding calcium-independent nitric oxide synthase confers inducibility by interferon-γ and bacterial lipoposaccharide. J Exp Med 177: 1779–1784

Xie QW, Kashiwabara Y, Nathan C (1994) Role of transcription factor NF-κB/Rel in the induction of nitric oxide synthase. J Biol Chem 269: 4705–4708

Diffusible Messengers and Intercellular Signaling: Locally Distributed Synaptic Potentiation in the Hippocampus

D.V. MADISON and E.M. SCHUMAN

Diffusible messengers like nitric oxide (NO) have been proposed to act as intercellular signals in the production of long term potentiation (LTP; for review, see SCHUMAN and MADISON 1993). The postsynaptic generation of such a diffusible messenger leads to the prediction that synapses in the vicinity of the messenger-producing synapse will also be influenced. We have tested this prediction directly by recording simultaneously from two postsynaptic CA1 neurons and recording synaptic strength in response to Schaffer collateral stimulation before and after LTP induction (by pairing 1 Hz stimulation with postsynaptic depolarization) in one neuron (see also BONHOEFER et al. 1989). In experiments where we impaled two pyramidal neurons in close proximity (132.9 ± 39.6 μM intersomatic distance, ISD), the paired cell exhibited robust LTP (186.0 ± 15.6 mean percentage of baseline) and the nearby neighboring neuron also exhibited a significant increase in synaptic strength (130.6 ± 9.5%; n=19). Postsynaptic injection of the NO synthase inhibitor L-Me-Arg into the paired cell blocked LTP production in that cell (95.8 +/– 6.9%) as well as in the nearby cell (102.2 ± 10.1%; ISD = 143.3 ± 43.6 μM; n=18). This spreading of potentiation was spatially restricted as it was not observed when the neighboring cell was > 500 μM away (paired cell LTP, 170.1 ± 11.8%; distant neighbor, 97.6 ± 0.7%; ISD = 674.0 ± 38.3 μM; n=15). Since previous findings showed that LTP can be blocked by various postsynaptic manipulations, we were interested in whether these processes are also important in the neighboring cell. We found that a combination of dialysis, Ca^{2+} chelators, and membrane hyperpolarization in the nearby neighboring cell blocked the spreading of enhancement usually observed (paired 195.7 ± 43.1; neighbor, 105.6 ± 24.7; n=8). These results suggest that the neighboring cell may play an active role in the enhancement described, either as a target site for a diffusible signal, or perhaps for the generation of an additional messenger(s).

No matter what the exact molecular mechanisms underlying the intercellular communication of potentiation, these data suggest that the formation of synaptic changes previously thought to be restricted to synapses onto a single cell, can also result in synaptic changes at nearby synapses (see also GALLY et al. 1980). As

Beckman Center, Department of Molecular and Cellular Physiology, School of Medicine, Stanford University Medical Center, Stanford, CA 94305–5426, USA such, diffusible signals may act as retrograde synaptic signals or alternately may serve the function of amplifying synaptic transmission at synapses in close proximity to the site of messenger generation. These results may help to understand processes both in the developing and adult brain. During development, locally distributed potentiation may be important in the formation of functional arrays of neurons analogous to anatomically segregated anatomical structures such as cortial columns (MILLER et al. 1989). Thus, distribution of potentiation could serve to induce temporary functional domains in regions where no such anatomical specialization exists. To the extent that neurons in anatomical proximity frequently share common output targets, this type of enhancement may also serve the general purpose of locally amplifying synaptic signals which underlie common neuronal functions.

References

Schuman EM, Madison DV (1993) Nitric oxide as an intercellular signal in long-term potentiation. Semin Neurosci 5: 207–215
Bonhoefer T, Staiger V, Aertsen A (1989) Synaptic plasticity in rat hippocampal slice cultures; local "Hebbian" conjunction of pre- and postsynaptic stimulation leads to distributed synaptic enhancement. Proc Natl Acad Sci USA 86: 8113–8117
Gally JA, Montague PR, Reeke GN, Edelman GM (1990) The NO hypothesis; possible effects of a short-lived, rapidly diffusible signal in the development of function of the nervous system. Proc Natl Acad Sci USA 87: 3547–3551
Miller KD, Keller JB, Stryker MP (1989) Ocular dominance column development analysis and stimulation. Science 245: 605–615

Nitric Oxide Formation in the Vascular Wall: Regulation and Functional Implications

R. Busse, I. Fleming, and V. B. Schini

1 Introduction

Numerous investigations have indicated that in both health and disease the short-lived radical nitric oxide (NO) is a key effector in the vascular system (Fig. 1). NO is generated by a five-electron oxidation of one of the terminal guanidino nitrogen atoms of L-arginine, catalyzed by NO synthases, and reaches the surrounding target cells by simple diffusion. The principal physiological source of NO in the vascular system is the endothelium, which constitutively expresses a NO synthase. The most important functions of endothelium-derived NO are the control of blood flow, and hence the supply of oxygen to organs, and the control of blood cell interaction with the vascular wall. Under certain pathophysiological conditions, endothelial cells are no longer the main source of NO. Indeed, in most types of vascular cells (e.g., vascular smooth muscle cells, macrophages, fibroblasts and endothelial cells) a NO producing pathway is induced following exposure to cytokines, such as interleukin-1β (IL-1β) and tumor necrosis factor-α (TNF-α), the levels of which are elevated in response to infection and injury. NO generated by this pathway in large amounts may account, at least in part, for the cytotoxic effect of macrophages and thus play a crucial role in host defense. The finding that NO can have such contrasting effects, i.e., to be protective and yet also cytotoxic, may be explained by the involvement of two NO-generating systems regulated by distinctly different mechanisms.

Zentrum der Physiologie, Klinikum der JWG-Universität, Theodor-Stern-Kai 7, 60590 Frankfurt/Main, Germany

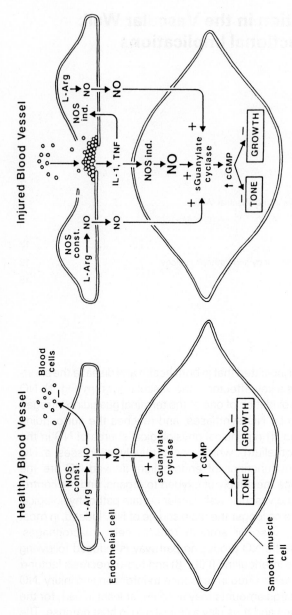

Fig. 1. In healthy blood vessels, nitric oxide (*NO*) is produced in the endothelial cells by the constitutive NO synthase (*NOS const.*). Destruction of the endothelium (e.g., by balloon catheterization) and exposure of the subendothelium rapidly results in the recruitment and activation of various blood cells. This inflammatory response is associated with the release of cytokines, such as interleukin-1 (*IL-1*) and tumor necrosis factor-α (*TNF-α*), which stimulate synthesis of the inducible NO synthase (*NOS ind.*) in the injured vessel wall

2 Nitric Oxide Biosynthetic Pathways in the Vascular Wall

2.1 Constitutive Nitric Oxide Synthase

The constitutive endothelial NO synthase displays binding sites for NADPH, flavin adenine dinucleotide, flavin mononucleotide and (6R)-5,6,7,8-tetrahydrobiopterin (BH_4) (LAMAS et al. 1992; SESSA et al. 1992; MARSDEN et al. 1992) and is strictly regulated by binding of the Ca^{2+}/calmodulin complex (BUSSE and MÜLSCH 1990a; FÖRSTERMANN et al. 1991). Calmodulin, in contrast with its role in other enzymes, has been shown to facilitate the reduction of the prosthetic heme group in NO synthase by transmitting electrons derived from NADPH (ABU-SOUD and STUEHR 1993).

NO synthase activity can be detected in the cytosol but activity is predominantly associated with the plasma membrane (HECKER et al. 1994; FÖRSTERMANN et al. 1991). The preferential association of the enzyme with membranes is likely to be a consequence of the myristoylation of the NH_2 terminal region of NO synthase (BUSCONI AND MICHEL 1993; SESSA et al. 1993). Localization of the enzyme in the plasma membrane may be of critical importance for the transduction mechanisms by which physical forces such as shear stress and pulsatile stretch elicit the formation of NO.

In situ, endothelial cells continuously produce low amounts of NO (GRIFFITH et al. 1984; MARTIN et al. 1986). This "basal" production can also be detected in cultured endothelial cells but decreases with increasing time in culture. It has recently been demonstrated that this phenomenon is paralleled by a decreased expression of the endothelial NO synthase (BUSSE et al. 1994). Expression of NO synthase mRNA and protein can, however, be up-regulated following the exposure of endothelial cell cultures to shear stress in vitro (BUSSE et al. 1994). Moreover a shear stress-responsive element in the promotor of the endothelial NO synthase gene has recently been identified, suggesting that a continuous physical stimulus such as shear stress transcriptionally regulates the expression of NO synthase in vivo (MARSDEN et al. 1992).

In addition to its basal release, both receptor-dependent and -independent mechanisms significantly enhance the release of NO from endothelial cells (for review see FURCHGOTT and VANHOUTTE 1989) (Fig. 2). Among the physiologically important receptor-dependent agonists are bradykinin, adenine nucleotides, 5-hydroxytryptamine (serotonin), substance P, and thrombin. These compounds activate specific receptors at the endothelial cell surface leading to a rapid increase in the intracellular concentration of free Ca^{2+} and, as a consequence of the Ca^{2+}/calmodulin dependency of the NO synthase, increase NO formation.

It is important to note, however, that the time course of the agonist-induced increase in NO synthase activity is significantly longer than that of the increase in intracellular Ca^{2+}. Thus, although a transient Ca^{2+} peak may be responsible for the initial increase in NO synthase activity, a second intracellular mechanism may be responsible for ensuring the maintained production of NO. Most of the

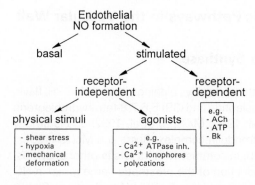

Fig. 2. The principal pathways by which nitric oxide (*NO*) can be formed in endothelial cells

physiologically relevant, receptor-dependent agonists activate the Na^+/H^+ exchanger leading to an increase in intracellular pH and it would appear that a sustained increase in NO formation can be observed following intracellular alkalinization. Indeed not only was the constitutive NO synthase found to be highly pH-sensitive in the physiological range but inhibition of the Na^+/H^+ exchanger significantly decreased bradykinin-stimulated NO formation (FLEMING et al. 1994).

2.2 Inducible Nitric Oxide Synthase(s)

An inducible NO synthase can be expressed in most types of vascular cells (see above) following their exposure to various inflammatory mediators such as cytokines (e.g., IL-1β, TNF-α) and bacterial lipopolysaccharides (for review see NATHAN 1992) (Table 1).

The inducible NO synthase is a predominantly cytosolic enzyme with essentially the same cofactor requirements as the constitutive enzyme. The major difference between this and the constitutive isoenzyme is that in the inducible NO synthase calmodulin is tightly bound in a Ca^{2+}-independent and noncovalent manner (CHO et al. 1992) and can be considered as an enzyme subunit, thus rendering its activity independent of changes in intracellular Ca^{2+}.

Table 1. Agents which influence expression of the inducible nitric oxide synthase

Inducers	Enhancers	Inhibitors
IL-1β	cAMP	TGFβ
TNF-α	EGF	$PDGF_{AB}$
IFN-γ	Basic FGF	$PDGF_{BB}$
LPS	Plasmin	IGF
		Thrombin

IL-1β, interleukin-1β; TGF-β, transforming growth factor-β; TNF-α, tumor necrosis factor-α; EGF, epidermal growth factor; PDGF, platelet-derived growth factor; IFN-γ, interferon-γ; FGF, fibroblast growth factor; IGF, insulin-like growth factor; LPS, bacterial lipopolysaccharide.

Once expressed, the enzyme is therefore always maximally activated. This observation may explain why the inducible L-arginine-NO pathway is associated with sustained production of large amounts of NO (IYENGAR et al. 1987)

The induction process requires a time lag of 4–6 h and involves activation of tyrosine kinases (MARCZIN et al. 1993; CORBETT et al. 1993), mRNA transcription and initiation of protein synthesis (for review see NATHAN 1992). In parallel with the initiation of inducible NO synthase synthesis, cytokines, by inducing the expression of guanosine triphosphate cyclohydrolase 1, also increase the de novo formation of the essential cofactor BH_4 (WERNER et al. 1989; FLEMING et al. 1991; GROSS and LEVI 1992). Inhibition of BH_4 synthesis prior to the exposure of vascular cells to cytokines completely prevents subsequent NO production, demonstrating that BH_4 synthesis is an absolute requirement for induction of NO synthesis.

Furthermore, NO synthase induction is facilitated by growth factors (such as basic fibroblast growth factor and epidermal growth factor, and the major fibrinolytic enzyme plasmin) which alone are without effect on inducible NO synthase expression (SCOTT-BURDEN et al. 1992; DURANTE et al. 1993). Increases in intracellular levels of cyclic adenosine 3',5'-monophosphate (cAMP) have a similar effect in that they enhance expression of the inducible NO synthase (KOIDE et al. 1993). The mechanisms underlying such an effect are unclear but may be due to activation of a cAMP regulatory element in the promotor of the inducible NO gene. It has, however, become apparent that the promotor of the gene encoding the murine inducible NO synthase does not contain a cAMP-responsive element (XIE et al. 1993), thus suggesting that cAMP exerts its effect by other, indirect, mechanisms. For example, cAMP has been reported to stabilize inducible NO synthase mRNA, prolonging its half life, rather than influencing its transcriptional activation (KUNZ et al. 1994).

The activation of many types of mammalian cells by a single cytokine and/or lipopolysaccharide is sufficient to cause the expression of the inducible NO synthase, but the combination of two or more stimuli leads to a synergistic enhancement of expression (KILBOURN and BELLONI 1990). Such findings suggest that the presence of threshold concentrations of several different cytokines at sites of vascular injury may result in the local generation of NO in amounts sufficient to elicit biological responses in neighboring cells. Although NO production by the inducible NO synthase is, at first glance, a process which is not subjected to any cellular regulation, it would appear that its expression can indeed be controlled by a variety of mechanisms. Platelets appear to play a significant role in the regulation of vascular NO formation, since substances released from aggregating platelets, such as platelet-derived growth factor, transforming growth factor-β and insulin-like growth factor, efficiently prevent expression of the inducible NO synthase in cultured vascular smooth muscle cells (SCHINI et al. 1992; DURANTE et al. 1994). In addition to these growth factors, the major coagulation protease thrombin prevents the induced formation of NO in vascular smooth muscle cells (SCHINI et al. 1993). Therefore, local activation of platelets and the coagulation cascade may prevent excessive formation of NO in vascular smooth muscle following injury, thus avoiding the cytotoxic effects of large amounts of NO.

3 Effector Pathways of Nitric Oxide

Nitric oxide is a multifunctional effector molecule in the vascular system which readily diffuses through lipid membranes and, in the presence of oxygen, is (within seconds) inactivated by conversion to nitrite and/or nitrate. Thus, the short half-life of NO constrains its biological action only to the generator cell itself and cells in its immediate vicinity. The major effector pathways mediating the action of NO in vascular cells have been summarized in Fig. 3.

The best characterized target for NO is the soluble guanylyl cyclase (CRAVEN and DeRUBERTIS 1978). NO binds to the iron present in the heme of the catalytic domain of the enzyme and by increasing its activity leads to an enhanced formation of cyclic guanosine 3',5'-monophosphate (KATSUKI et al. 1977). This biochemical pathway accounts for the vasodilatory properties of NO in all types of blood vessels studied to date and also for its ability to prevent platelet activation (GRUETTER et al. 1979; MELLION et al. 1981). These responses are eventually due to an attenuation of Ca^{2+} signaling (BUSSE et al. 1987; KARAKI et al. 1988).

NO can also react with non-heme iron in target cells and such a reaction has been involved in host defense against infection. Cytotoxic activated macrophages, by releasing large amounts of NO, precipitate the loss of intracellular iron resulting in the inhibition of certain vital enzymes which require iron as an essential cofactor. Such enzymes include the NADH: ubiquinone oxidoreductase and succinate: ubiquinone oxidoreductase (both of which are involved in mitochondrial respiration), the critic acid cycle enzyme aconitase and ribonucleotide reductase (the rate limiting enzyme in DNA replication) (DRAPIER and HIBBS 1986; HIBBS et al. 1988).

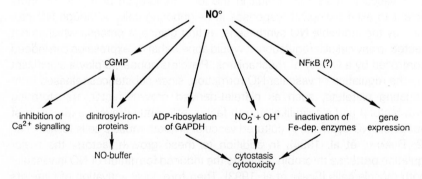

Fig. 3. The effects of nitric oxide (NO). NFκB, nuclear factor κB; Fe dep, non-heme iron-dependent enzymes

Other important cellular targets for NO are sulfhydryl (SH)-containing proteins. NO derivatives, such as the nitrosonium ion (NO⁺), interact with these SH groups to produce biologically active S-nitrosoproteins (STAMLER et al. 1992b) (see following chapter by J.S. Stamler, pp 19–36). S-nitrosylation appears to be implicated in both the beneficial and non beneficial actions of NO. Among the thiol-containing proteins of potential physiological significance for S-nitrosylation are serum albumin and tissue-type plasminogen activator (STAMLER et al. 1992a,b). These S-nitrosylated proteins, like NO, are potent vasodilators and inhibitors of platelet aggregation, effects mediated via the activation of soluble guanylyl cyclase in the target cells. However, the half-lives of these compounds are significantly longer (>30 min) than that of NO (STAMLER et al. 1992b; KEANEY et al. 1993). These NO-containing compounds may therefore serve as adducts, which by stabilizing the highly reactive NO prolong its activity and/or facilitate its biological action thus extending its sphere of influence. In addition, NO, by S-nitrosylating the glycolytic enzyme glyceraldehyde-3-phosphate dehydrogenase decreases its activity and increases its auto-ADP-ribosylation (MOLINA-Y-VEDIA et al. 1992; DIMMELER et al. 1992). The impairment of this glycolytic enzyme by NO may, in association with the inhibition of iron-sulfur enzymes like aconitase and electron transport proteins of the respiratory chain (see above), contribute to the cytotoxic effects of NO.

Furthermore, NO may also modulate vascular functions by controlling the expression of genes encoding certain vasoactive proteins. Indeed, in cultured endothelial cells NO has been shown to down-regulate the expression of the chemoattractant MCP-1 (monocyte chemoattractant protein) (SCHRAY-UTZ et al. 1993), the potent vasoconstrictor endothelin-1, and the chemoattractant and vascular smooth muscle growth factor platelet derived growth factor (KOUREMBANAS et al. 1993). The demonstration that NO is able to modulate the expression of certain genes underlines the significance of endothelium-derived NO, both in short-term and long-term vascular homeostasis.

4 Role of Nitric Oxide in Vascular Physiology and Pathophysiology

In humans and experimental animals, the administration of structural analogues of L-arginine, which competitively inhibit NO synthase, results in a rapid and sustained increase in arterial blood pressure and a decrease in blood flow (REES et al. 1989a; AISAKA et al. 1989; VALLANCE et al. 1989). This pressor effect, which is associated with the constriction of several peripheral vascular beds, can be reversed by the infusion of L-arginine (REES et al. 1989a; AISAKA et al. 1989; GARDINER et al. 1990; GALLE et al. 1993) and is a consequence of the inhibition of basally released endothelium-derived NO. The augmentation of contractile responses in isolated preconstricted arteries following removal of the endo-thelium or incubation with NO synthase inhibitors suggests that vascular tone is a net effect of constrictor and dilator influences (REES et al. 1989b; MÜLSCH and BUSSE 1990; MOORE et al. 1990). Taken together, these observations indicate that NO derived from endothelial cells regulates the tone of the underlying vascular smooth muscle, and hence contributes to the control of regional blood flow. The most important physiological stimulus for the basal release of NO from the endothelium is the fluid shear stress imposed on the luminal surface of endothelial cells by the streaming blood. Indeed, in vitro investigations in the intact coronary vascular bed and in arterial segments under biossay conditions have shown that an increase in fluid shear stress triggers the release of endothelium-derived NO (HECKER et al. 1993; LAMONTAGNE et al. 1992) (Fig. 4).

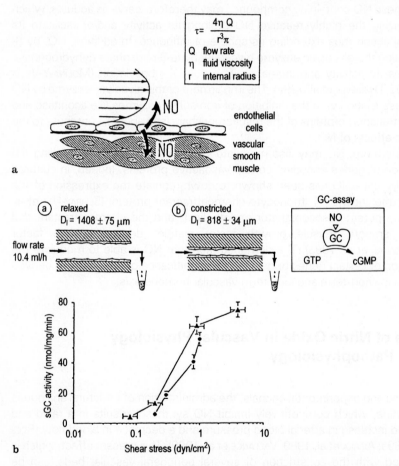

Fig. 4a, b. The relationship between fluid shear stress (τ) and the endothelial cell surface. Shear stress in rabbit femoral artery segments was increased either by **a** increasing flow through the vessel or **b** by decreasing vessel diameter. The graph shows the relationship between increases in shear elicited by vesoconstriction (triangles) or increased flow (circles) and nitric oxide (*NO*) release from rabbit femoral artery segments. NO production was determined by stimulation of soluble guanylyl cyclase (*sGC*) and expressed as net increase in enzyme activity over basal values. Shear was estimated under the simplifying assumption of a parabolic velocity profile

Shear stress-stimulated NO formation by endothelial cells, which is associated with an enhanced expression of the endothelial NO synthase, may contribute to the beneficial effects of physical exercise on the cardiovascular system (NISHIDA et al. 1992; SESSA et al. 1994). The physiologial significance of this enhanced formation of NO may be considered as a means of acutely increasing local blood flow, and thus supply of oxygen, in response to the greater metabolic requirements of tissues under working conditions (POHL et al. 1994). In addition to this "fundamentally vital" basal release of endothelium-derived NO, its formation is enhanced by physiologically relevant mediators such as hormones, neurotransmitters, and platelet-derived substances (see above). This additional release may play a significant role in the prevention of excessive vasoconstriction and abnormal coagulation following trauma to the blood vessel wall. NO may also contribute to the regulation of vascular growth and the remodeling of the blood vessel wall following trauma, since low levels of NO have been shown to effectively inhibit the proliferation of cultured vascular smooth muscle cells (GARG and HASSID 1989; SCOTT-BURDEN et al. 1992).

Since NO exerts a variety of effects on vascular function, it follows that a reduction as well as an increase in vascular NO formation has wide ranging consequences. For example, a decrease in NO production by a dysfunctional endothelium has been implicated in the pathogenesis of hypertension. The loss of the antiproliferative effect of NO may contribute to the vascular hypertrophy which characterizes a number of pathological states including atherosclerosis (for review see BUSSE and FLEMING 1993). An overproduction of NO, such as that observed during endotoxic shock, following the induction of the inducible NO synthase in vascular smooth muscle cells (BUSSE and MÜLSCH 1990b), may at first glance be desirable for defense against the invading pathogen. However, it eventually results in an unrelenting hypotension and hyporeacivity to vasoconstrictor agents (THIEMERMANN and VANE 1990; JULOU-SCHAEFFER et al. 1991).

In summary, it appears that the release of endothelium-derived NO, a potent antagonistic agent counteracting the development of local vasospasm, coagulation and aberrant proliferation of the underlying smooth muscle cells, plays an essential role in the short- and long-term homeostasis of the blood vessel wall. The induction of nonendothelial NO forming pathways in the vascular wall may be Janus-like, in that whereas an excessive and uncontrolled widespread production within the vasculature may be lethal, a more confined induction of NO synthase at sites of endothelial injury may be beneficial and compensate for the loss of endothelium-derived NO.

References

Abu-Soud HM, Stuehr DJ (1993) Nitric oxide synthases reveal a role for calmodulin in controlling electron transfer. Proc Natl Acad Sci USA 90: 10769–10772

Aisaka K, Gross SS, Griffith OW, Levi R (1989) N^G-methylarginine, an inhibitor of endothelium-derived nitric oxide synthesis, is a potent pressor agent in the guinea pig – does nitric oxide regulate blood pressure in vivo. Biochem Biophys Res Commun 160: 881–886

Busconi L, Michel T (1993) Endothelial nitric oxide synthase: N-terminal myristoylation determines subcellular localization. J Biol Chem 268: 8410–8413

Busse R, Fleming I (1993) The endothelial organ. Curr Opin Cardiol 8: 719–727

Busse R, Mülsch A (1990a) Calcium-dependent nitric oxide synthesis in endothelial cytosol is mediated by calmodulin. FEBS Lett 265: 133–136

Busse R, Mülsch A (1990b) Induction of nitric oxide synthase by cytokines in vascular smooth muscle cells. FEBS Lett 275: 87–90

Busse R, Lückhoff A, Bassenge E (1987) Endothelium-derived relaxant factor inhibits platelet activation. Naunyn-Schmiedebergs Arch Pharmacol 336: 566–571

Busse R, Hecker M, Fleming I (1994) Control of nitric oxide and prostacyclin synthesis in endothelial cells. Arzneimittel Forschung/Drug Res 44[Suppl]: 392–396

Cho HJ, Xie QW, Calaycay J, Mumford RA, Swiderek KM, Lee TD, Nathan C (1992) Calmodulin is a subunit of nitric oxide synthase from macrophages. J Exp Med 176: 599–604

Corbett JA, Sweetland MA, Lancaster JR, McDaniel ML (1993) A 1-hour pulse with IL-1β induces formation of nitric oxide and inhibits insulin secretion by rat islets of Langerhans: evidence for a tyrosine kinase signaling mechanism. FASEB J 7: 369–374

Craven PA, DeRubertis FR (1978) Restoration of the resposiveness of purified guanylyl cyclase to nitrosoguanidine, nitric oxide, and related activators ba heme and hemeproteins. J Biol Chem 253: 8433–8443

Dimmeler S, Lottspeich F, Brune B (1992) Nitric oxide causes ADP-ribosylation and inhibition of glyceraldehyde-3-phosphate dehydrogenase. J Biol Chem 267: 16771–16774

Drapier JC, Hibbs JB (1986) Murine cytotoxic activated macrophages inhibit aconitase in tumor cells. Inhibition involves the iron-sulfur prosthetic group and is reversible. J Clin Invest 78: 790–797

Durante W, Schini VB, Catovsky S, Kroll MH, Vanhoutte PM, Schafer AI (1993) Plasmin potentiates induction of nitric oxide synthesis by interleukin-1β in vascular smooth muscle cells. Am J Physiol 264: H617–H624

Durante W, Schini VB, Kroll MH, Catovsky S, Scott-Burden T, White JG, Vanhoutte PM, Schafer AI (1994) Platelets inhibit the induction of nitric oxide synthesis by interleukin-1β in vascular smooth muscle cells. Blood 83: 1831–1838

Fleming I, Gray GA, Schott C, Stoclet JC (1991) Inducible but not constitutive production of nitric oxide by vascular smooth muscle cells. Eur J Pharmacol 200: 375–376

Fleming I, Hecker M, Busse R (1994) Intracellular alkalinization induced by bradykinin sustains activation of the constitutive nitric oxide synthase in endothelial cells. Circ Res 74: 1220–1226

Förstermann U, Pollock JS, Schmidt HHHW, Heller M, Murad F (1991) Calmodulin-dependent endothelium-derived relaxing factor/nitric oxide synthase activity is present in the particulate and cytosolic fractions of bovine aortic endothelial cells. Proc Natl Acad Sci USA 88: 1788–1792

Furchgott Rf, Vanhoutte PM (1989) Endothelium-derived relaxing and contracting factors. FASEB J 3: 2007–2018

Galle J, Bauersach J, Bassenge E, Busse R (1993) Arterial size determines the enhancement of contractile responses after supression of endothelium-derived relaxing factor formation. Pflugers Arch 422: 564–569

Gardiner SM, Compton AM, Bennett T, Palmer RMJ, Moncada S (1990) Control of regional blood flow by endothelium-derived nitric oxide. Hypertension 15: 486–492

Garg UC, Hassid A (1989) Nitric oxide-generating vasodilators and 8-bromo-cyclic guanosine monophosphate inhibit mitogenesis and proliferation of cultured rat vascular smooth muscle cells. J Clin Invest 83: 1774–1777

Griffith TM, Edwards DH, Lewis MJ, Newby AC, Henderson AH (1984) The nature of endothelium-derived vascular relaxant factor. Nature 308: 645–647

Gross SS, Levi R (1992) Tetrahydrobiopterin synthesis. J Biol Chem 267: 25722–25729

Gruetter CA, Barry BK, McNamara BD, Gruetter DY, Kadowitz PJ, Ignarro LJ (1979) Relaxation of bovine coronary artery and activation of coronary arterial guanylate cyclase by nitric oxide, nitroprusside and a carcinogenic nitrosoamine. J Cycl Nucl Res 5: 211–224

Hecker M, Mülsch A, Bassenge E, Busse R (1993) Vasoconstriction and increased flow: two principle mechanisms of shear stress-dependent endothelial autacoid relase. Am J Physiol 265: H828–H833

Hecker M, Mülsch A, Bassenge E, Förstermann U, Busse R (1994) Subcellular localization and characterization of nitric oxide synthase(s) in endothelial cells - Physiologic implications. Biochem J 299: 247–252

Hibbs JB, Taintor RR, Vavrin Z, Rachlin EM (1988) Nitric oxide: a cytotoxic activated macrophage effector molecule. Biochem Biophys Res Commun 157: 87–94

Iyengar R, Stuehr DJ, Marletta MA (1987) Macrophage synthesis of nitrite, nitrate, and N-nitrosamines: precursors and role of the respiratory burst. Proc Natl Acad Sci USA 84: 6369–6373

Julou-Schaeffer G, Gray GA, Fleming I, Parratt JR, Stoclet JC (1991) Activation of the L-arginine pathway is involved in vascular hyporeactivity induced by endotoxin. J Cardiovasc Pharmacol 17: S207–S212

Karaki H, Sato K, Ozaki H, Murakami K (1988) Effects of sodium nitroprusside on cytosolic calcium level in vascular smooth muscle. Eur J Pharmacol 156: 259–266

Katsuki S, Arnold WP, Mittal CK, Murad F (1977) Stimulation of guanylate cyclase by sodium nitroprusside, nitroglycerin and nitric oxide in various tissue preparations and comparison to the effects of sodium azide and hydroxylamine. J Cycl Nucl Prot Phosphoryl Res 3: 23–35

Keaney JF, Simon DI, Stamler JS, Jaraki O, Scharfstein J, Vita JA, Loscalzo J (1993) NO forms an adduct with serum albumin that has endothelium-derived relaxing factor-like properties. J Clin Invest 91: 1582–1589

Kilbourn RG, Belloni P (1990) Endothelial cell production of nitrogen oxides in response to interferon-gamma in combination with tumor necrosis factor, interleukin-1, or endotoxin. J Natl Cancer Ins 82: 772–776

Koide M, Kawahara Y, Nakayama I, Tsuda T, Yokoyama M (1993) Cyclic AMP elevating agents induce an inducible type of nitric oxide synthase in cultured vascular smooth muscle cells. J Biol Chem 268: 24959–24966

Kourembanas S, McQuillan LP, Leung GK, Faller DV (1993) Nitric oxide regulates the expression of vasoconstrictors and growth factors by vascular endothelium under both normoxia and hypoxia. J Clin Invest 92: 99–104

Kunz D, Mühl H, Walker G, Pfeilschifter J (1994) Two distinct signalling pathways trigger the expression of inducible nitric oxide synthase in rat mesangial cells. Proc Natl Acad Sci USA 91: 5387–5391

Lamas S, Marsden PA, Li GK, Tempst P, Michel T (1992) Endothelial nitric oxide syntahse: molecular cloning and characterization of a distinct constitutive enzyme isoform. Proc Natl Acad Sci USA 89: 6348–6352

Lamontagne D, Pohl U, Busse R (1992) Mechanical deformation of vessel wall and shear stress determine the basal release of endothelium- derived relaxing factor in the intact rabbit coronary vascular bed. Circ Res 70: 123–130

Marczin N, Papapetropoulos A, Catravas JD (1993) Tyrosine kinase inhibitors suppress endotoxin- and IL-1β-induced NO synthesis in aortic smooth muscle cells. Am J Physiol 265: H1014–H1018

Marsden PA, Schappert KT, Chen HS, Flowers M, Sundell CI, Wilcox JN, Lamas S, Michel T (1992) Molecular cloning and characterization of human endothelial nitric oxide synthase. FEBS Lett 307: 287–293

Martin W, Furchgott RF, Villani GM, Jothianandan D (1986) Depression of contractile responses in rat aorta by spontaneously released endothelium-derived relaxing factor. J Pharmacol Exp Ther 237: 529–538

Mellion BT, Ignarro LJ, Ohlstein EH, Pontecorvo EG, Hyman AL, Kadowitz PJ (1981) Evidence for the inhibitory role of guanosine 3',5'-monophosphate in ADP-induced human platelet aggregation in the presence of nitric oxide and related vasodilators. Blood 57: 946–955

Molina-y-Vedia L, McDonald B, Reep B, Brüne B, DiSilvio M, Billiar TR, Lapetina EG (1992) Nitric oxide-induced S-nitrosylation of glyceraldehyde-3-phosphate dehydrogenase inhibits enzymatic activity and increases endogenous ADP-ribosylation. J Biol Chem 267: 24929–24932

Moore PK, Alswayeh OA, Chong NWS, Evans RA, Gibson A (1990) L-N^G-nitro arginine (L-NOARG), a novel L-arginine-reversible inhibitor of endothelium-dependent vasodilatation in vitro. Br J Pharmacol 99: 408–412

Mülsch A, Busse R (1990) N^G-nitro-L-arginine (N^δ-[imino(nitroamino)methyl]-L-ornithine) impairs endothelium-dependent dilations by inhibiting cytosolic nitric oxide synthesis from L-arginine. Naunyn-Schmiedebergs Arch Pharmacol 341: 143–147

Nathan C (1992) Nitric oxide as a secretory product of mammalian cells. FASEB J 6: 3051–3064

Nishida K, Harrison DG, Navas JP, Fisher AA, Dockery Sp, Uematsu M, Nerem RM, Alexander RW, Murphy TJ (1992) Molecular cloning and characterization of the constitutive bovine aortic endothelial cell nitric oxide synthase. J Clin Invest 90: 2092–2096

Pohl U, Lamontagne D, Bassenge E, Busse R (1994) Attenuation of coronary autoregulation in the isolated rabbit heart by endothelium derived nitric oxide. Cardiovasc Res 28: 414–419

Rees DD, Palmer RM, Moncada S (1989a) Role of endothelium-derived nitric oxide in the regulation of blood pressure. Proc Natl Acad Sci USA 86: 3375–3378

Rees DD, Palmer RMJ, Hodson HF, Moncada S (1989b) A specific inhibitor of nitric oxide formation from L-arginine attenuates endothelium-dependent relaxation. Br J Pharmacol 96: 418–424

Schini VB, Durante W, Elizondo E, Scott-Burden T, Junquero DC, Schafer AI, Vanhoutte PM (1992) The induction of nitric oxide synthase activity is inhibited by TGF-β1, PDGF$_{AB}$ and PDGF$_{BB}$ in vascular smooth muscle cells. Eur J Pharmacol 216: 379–383

Schini VB, Catovsky S, Durante W, Scott-Burden T, Schafer AI, Vanhoutte PM (1993) Thrombin inhibits induction of nitric oxide synthase in vascular smooth muscle cells. Am J Physiol 264: H611–H616

Schray-Utz B, Zeiher AM, Busse R (1993) The expression of monocyte chemoattractant protein (MCP-1) mRNA in human endothelial cells is modulated by nitric oxide. FASEB J 7: A130

Scott-Burden T, Schini VB, Elizondo E, Junquero DC, Vanhoutte PM (1992) Platelet-derived growth factor suppresses and fibroblast growth factor enhances cytokine-induced production of nitric oxide by cultured smooth muscle cells; effects on cell proliferation. Circ Res 71: 1088–1100

Sessa WC, Harrison JK, Barber CM, Zeng D, Durieux ME, D' Angelo DD, Lynch KE, Peach MJ (1992) Molecular cloning and expression of cDNA encoding endothelial cell nitric oxide synthase. J Biol Chem 267: 15274–15276

Sessa WC, Barber CM, Lynch KR (1993) Mutation of N-myristoylation site converts endothelial cell nitric oxide synthase from a membrane to a cytosolic protein. Circ Res 72: 921–924

Sessa WC, Pritchard K, Seyedi N, Wang J, Hintze TH (1994) Chronic exercise in dogs increases coronary vascular nitric oxide production and endothelial cell nitric oxide synthase gene expression. Circ Res 74: 249–353

Stamler JS, Jaraki O, Osborne J, Simon DI, Keaney J, Vita J, Singel D, Valeri CR, Loscalzo J (1992a) Nitric oxide circulates in mammalian plasma primarily as an S-nitroso adduct of serum albumin. Proc Natl Acad Sci USA 89: 7674–7677

Stamler JS, Simon DI, Osborne JA, Mullins ME, Jaraki O, Michel T, Singel DJ, Loscalzo J (1992b) S-nitrosylation of proteins with nitric oxide—synthesis and characterization of biologically active compounds. Proc Natl Acad Sci USA 89: 444–448

Thiemermann C, Vane JR (1990) Inhibition of nitric oxide synthesis reduces the hypotension induced by bacterial lipopolysaccharides in the rat in vivo. Eur J Pharmacol 182: 591–595

Vallance P, Collier J, Moncada S (1989) Effects of endothelium-derived nitric oxide on peripheral arteriolar tone in man. Lancet 2: 997–1000

Werner ER, Werner-Felmayer G, Fuchs D, Hausen A, Reibnegger G, Wachter H (1989) Parallel induction of tetrahydrobiopterin biosynthesis and indoleamine 2,3-dioxygenase activity in human cells and cell lines by interferon-gamma. Biochem J 262: 861–868

Xie Qw, Whisnant R, Nathan C (1993) Promotor of the mouse gene encoding calcium-dependent nitric oxide synthase confers inducibility by interferon gamma and bacterial lipopolysaccharide. J Exp Med 177: 1779–1784

S-Nitrosothiols and the Bioregulatory Actions of Nitrogen Oxides Through Reactions with Thiol Groups

J.S. STAMLER

1 History and Overview

Chemists have long been aware of the red color that develops upon treatment of thiols with nitrous acid. Shortly after the turn of the last century, TASKER and JONES (1909) reported on the synthesis of benzene thionitrite, which exhibits a red color. The authors further noted that the compound was highly unstable and rapidly decomposed to (biphenyl) disulfide and nitric oxide (NO˙) gas. Thermal and photolytic decomposition of thionitrites was later shown to involve homolytic fission, as inferred from these early experiments (LECHER and SIEFKEN 1926; RAO et al. 1967; JOSEPHY et al. 1984). TASKER and JONES (1909) also described the thionitrite (or S-nitrosothiol; RS-NO) formed from ethane-thiol treatment with nitrosyl chloride (NOCl). This compound was shown to be significantly more stable than the corresponding benzene thiol derivative, but also disappeared with evolution of nitric oxide. Thus, the well documented importance of the electron withdrawing effect of the thiyl (RS) group in hastening the homolytic decomposition of RS-NO had been appreciated well over 50 years ago. In 1969, MIRNA and HOFMAN provided additional insight into the physical properties of biological RS-NOs. These studies demonstrated the trend for greater stability of thionitrites at low pH. At the same time, differences in the stability of thionitrites derived from cysteine and

Divisions of Respiratory Medicine and Cardiovascular Medicine, Duke University Medical Center, Bell Building, Room 350, Durham, NC 27710, USA glutathione were noted. While S-nitroso-cysteine rapidly decomposes through homolytic fission, the S-nitroso adduct of glutathione remains stable over a wide (physiological) pH range (MIRNA and HOFMANN 1969). Shortly thereafter, FIELD and colleagues, isolated the highly stable thionitrite derivative of N-acetylpenicillamine (FIELD et al. 1978). More importantly, this work also demonstrated that disappearance of RS-NO can follow heterolytic pathways, specifically, reactions in which RS-NO formally transfers NO^+ (or NO^-). Additional reactions, persumed to be heterolytic in mechanism, were subsequently reported by MASSEY and colleagues (1978) and OAE and coworkers (1978) and supported the growing use of thionitrites in organic synthesis as effective nitrosating agents. The notable stability of protein thionitrites has been appreciated most recently, and heterolytic fission of the S-N bond appears to predominate in many biological systems (STAMLER et al. 1992a,b,c; LIPTON et al. 1993; STAMLER 1994).

In a series of papers in the early 1980s, IGNARRO and coworkers demonstrated the smooth muscle relaxant and antiplatelet properties of RS-NO (IGNARRO and GRUETTER 1980; GRUETTER et al. 1981; IGNARRO et al. 1981). These newer biochemical functions were attributed to activation of the enzyme guanylate cyclase, which catalyzes the transformation of guanosine triphosphate (GTP) to cyclic guanosine monophosphate (cGMP) (IGNARRO and GRUETTER 1980; GRUETTER et al. 1981). The later discovery that cyclic GMP (cGMP) acted as a signal transduction mechanism for the L-arginine: NO˙ pathway in mammalian cells, in which NO˙ is conceived to activate guanylate cyclase by interaction with its heme center, strengthened the resolve that nitric oxide (NO˙) itself is the bioactive moiety in RS-NO (IGNARRO 1989). This line of reasoning, taken together with an early focus on the biochemistry of S-nitroso-cysteine, which readily undergoes homolysis to yield NO˙ in vitro (MIRNA and HOFMANN 1969; KANNER 1979; MYERS et al. 1990), led to the common view that all RS-NOs are inherently unstable molecules which decompose to liberate nitric oxide. It is clear from the above cited literature, however, that many of these recent views do not encompass the larger chemistry of thionitrites. Moreover, as will become more evident, the (rapid) homolytic liberation of NO˙ by RS-NO is, in part, an artifact of in vitro systems containing contaminant metals, which mediate this pathway. In this perspective I will attempt to integrate the broader chemistry of RS-NO in a review of their biological properties in order to explain their diverse biological actions.

2 Physical Properties and Chemistry of S-Nitrosothiols

S-nitroso derivatives of primary aminothiols are usually red (IGNARRO et al. 1980; OAE and SHINHAMA 1983; STAMLER and LOSCALZO 1992). Tertiary thionitrites are green colored (FIELD et al. 1978; OAE and SHINHAMA 1983). Solutions of protein RS-NOs are orange or rose colored (STAMLER et al. 1992a). The stability of thionitrites varies as a function of the thiyl group (RS). With an increase in electron withdrawing effect of R, or the proximity of a primary amino group, the stability of RS-NO tends to decrease (TASKER and JONES 1909; MORRIS and HANSEN 1981; J.S. Stamler, personal observation). This may be a consequence of a change in thiol pK incurred by the adjacent group (in which heterolytic fission is governed by the basicity of RS⁻), or the role the amino group may play in catalyzing the decomposition of RS-NO by way of a thionitrite to nitosamine rearrangement. Recent studies suggest that contaminant metals, and in particular copper ions, facilitate thionitrite decomposition (RAMDEV et al. 1993; MCNAINLY and WILLIAMS 1993); this affect may involve the intermediate formation of Cu-complexes (MCNAINLY and WILLAMS 1993), which are almost certain to involve copper amine coordination. The finding that moderate acidity (pH 1–5) markedly increases the stability of RS-NO may be explained, in part, by the pH limiting affects on metal complexation. However, RS-NOs exhibit very complex pH-dependent stability profiles in which acid catalyzed hydrolysis of RS-NO is balanced by product (NO⁺) mediated *S*-nitrosation (Eq. 1), and alkaline conditions lend to heterolytic fission (Eq. 2)

$$RS\text{-}NO + H^+ \rightleftarrows RSH + NO^+, \tag{1}$$

$$RS\text{-}NO + OH^- \rightarrow NO_2^- + RS^- + H^+. \tag{2}$$

The importance of metals in decomposition reactions may also rationalize the much higher stability of homocysteine thionitrite than of its cysteine homologue (STAMLER et al. 1993; RAMDEV et al. 1993), as the sulfur is distanced from the terminal amine by one extra methylene, perhaps limiting Cu²⁺ coordination. Further, modification of the amino group of cysteine by *N*-acetylation, dramatically stabilizes the derivative RS-NO (WILLIAMS 1988); in this case, however, internal stabilization may also derive from ring isomerization.

Bulky protecting groups also stabilize thionitrites. Examples include the thionitrite derivatives of *N*-acetyl-penicillamine and *t*-butyl-thionitrite (FIELD et al. 1978; OAE and SHINHAMA 1983). Further, protein RS-NOs are often more stable than the *S*-nitroso derivatives of amino acids and small peptides (STAMLER et al. 1992a). In the latter case, the local microenvironment of the protein thiol may stabilize the thionitrite by supporting an acidic millieu, through intramolecular interactions and/or by limiting the access of reactive nucleophiles. Although it may be easy to envision unstable protein nitrosothiols, those tested to date, i.e. derivatives of albumin (STAMLER et al. 1992a), tissue plasminogen activator (STAMLER et al. 1992a,d,e), cathepsin B (STAMLER et al. 1992a,e), and soybean extract (MASSEY et al. 1978) form comparatively stable *S*-nitroso derivatives at physiologic pH. Disproportionation of RS-NO is accelerated by O_2 and slowed by excess thiol (RSH); not only is the latter true for highly stable thionitrites such as S-nitroso-glutathione but also for unstable compounds such as S-nitroso-cysteine (RAMDEV et al. 1993). Thiols may stabilize RS-NO by scavenging reactive oxygen species and/or complexing metals, to limit their respective roles in thionitrite disproportionation. Indeed, the critical importance of (contaminant) metals in the mechanism of homolytic fission of RS-NO (RAMDEV et al. 1993; MCNAINLY and WILLIAMS 1993), taken with the relative resistance of thionitrites to hydrolysis (STAMLER et al. 1992c), indicates that the apparent "instability" of these species in vitro is largely artifactual: truly physiological systems contain millimolar concentrations of thiol and free metals are sparse.

RS-NOs exhibit ultraviolet absorption peaks at approximately 200 nm (e ≈ 10 000 M^{-1} cm^{-1}), 335 nm (e ≈ 1000 M^{-1} cm^{-1}), approximately 500 nm (e ≈ 5 M^{-1} cm^{-1}), and 550 nm (e ≈ 5 M^{-1} cm^{-1}), (OAE and SHINHAMA 1983; BYLER et al. 1983; STAMLER et al. 1993). The extinction coefficients of protein RS-NO at ≈ 330 nm and 550 nm are on the order to three fold greater than RS-NOs (STAMLER et al. 1992a,d). The absorption maxima at ≈ 500 and 550 nm are unique to S-nitrosothiols while alkyl nitrites and nitrosoamines absorb in the lower ranges as well. Thionitrites possess infrared NO stretching frequencies in the range of 1490–1700 cm^{-1} and bending frequencies in the ranges of 610–660 cm^{-1} (OAE et al. 1978; OAE and SHINHAMA 1983; MASON 1969). The corresponding alkyl nitrites exhibit NO stretching vibrations at smaller wavelengths, which may be explained by the greater electronegativity of oxygen vs sulfur. ^{15}N-NMR spectra show shifts for RS-NO in the 750–790 ppm range as compared against an internal ^{15}N natural abundance spectrum of $NaNO_2$ at 587 ppm (BONNETT et al. 1975). Protein RS-NO exhibit ^{15}N-NMR shifts in the same range (STAMLER et al. 1992a,d).

The synthesis of thionitrites most commonly involves S-nitrosation chemistry (i.e., involving reaction or transfer of NO^+, the nitrosonium ion) (ALDRED and WILLIAMS 1982; STAMLER et al. 1992c)

$$RSH + XNO \rightleftarrows RSNO + HX.$$

The synthesis has been best characterized in acidic aqueous solution (pH ≤ 1) with nitrous acid (HNO_2) (STAMLER et al. 1992a,d; WILLIAMS 1988; STAMLER and LOSCALZO 1992); however, treatment of thiols with N_2O_4 in inert solvents such as $CHCl_3$ or CCl_4 is a very convenient alternative method (OAE et al. 1977). Nitrosation has also been carried out with NOCl, alkyl nitrites (RONO) and dinitrogen trioxide (N_2O_3) (OAE and SHINHAMA 1983). Nitrosation of thiols with nitric oxide gas requires the presence of oxygen in vitro (PRYOR et al. 1982). This reaction is presumed to occur by way of reactive intermediates such as N_2O_3 and N_2O_4 (PRYOR et al. 1982; STAMLER et al. 1992c; WINK et al. 1993) and likely serves as one pathway for RS-NO formation in vivo. Direct S-nitrosation of thiols by metal (Fe or Cu) nitrosyl compounds has also been reported and is consistent with an eletrophilic attack by NO^+, donated by the metal nitrosyl compound (STAMLER et al. 1992c). Metal nitrosyl form effectively under neutral physiological conditions in the presence of nitric oxide, and may account, in part, for the presence of·

thionitrites in biological systems (STAMLER et al. 1992c). Nitrosoamines also fall into the catagory of biological nitrosonium (NO^+) equivalents (STAMLER et al. 1992c) and directly effect S-nitrosation, as exemplified in the well studied case of thiourea (WILLIAMS 1988

$$R_2NNO + (NH_2)_2CS \xrightarrow{H^+} (NH_2)_2 CS^+NO + R_2NH.$$

Thionitrites may also form in a radical reaction between nitric oxide and thiyl radical (RS·) (WILLIAMS 1988). There is one report of RS-NO formation in the reaction involving nitroxyl (NO^-) and disulfide (RINDEN et al. 1989); however, the mechanistic details have not been worked out. There is also evidence for thiol nitrosation mediated by peroxynitrite ($OONO^-$) (Wu et al. 1994; MOHR et al. 1994), which is produced in the diffusion-limited reaction of NO˙ and superoxide. This pathway may help rationalize the high concentrations of RS-NO produced with inflammation (GASTON et al. 1993).

The kinetics of thionitrite formation are reasonably well characterized in the reaction between RSH and nitrite in acid solutions (BYLER et al. 1983; ALDRED and WILLIAMS 1982; AL-KAABI et al. 1982). This reaction exhibits a first order dependence on thiol, nitrous acid, and H^+. The forward reaction is interpreted as one involving S-nitrosation by NO^+ (or H_2ONO^+) at pH less than 1, and N_2O_3 at pH of greater than 2 (ALDRED and WILLIAMS 1982; RIDD 1978; WILLIAMS 1988). The rate expression for this reaction is as follows:

Rate = k_2 [HNO_2] [H^+] [RSH].

Much less is known about the mechanistic details of nitrosation reactions in biological systems. The transnitrosation reactions afforded by thionitrites, including those with secondary amines, alkylamides and alcohols, have not been studied mechanistically; however, it is likely that the nitrosation occurs directly and not by way of alternative nitroso intermediates (ALDRED and WILLIAMS 1982).

Thionitrites can disproportionate both homolytically and heterolytically. For example, both low molecular weight and protein thiol may participate in transnitrosations (reaction or transfer of NO^+) with other thionitrites (SIMON et al. 1993; BUTLER and ASKEW 1993; PARK 1988; STAMLER et al. 1992c). Moreover, some hydrolysis of thionitrites is to be expected under the acidic conditions found in lysosomal compartments of the gastrointestinal tract. That such reactions could lead to formation of carcinogenic nitrosoamines has been a source of some concern. Alternatively, very high excess of thiol (or vicinal thiols on proteins) may promote the liberation of nitroxyl anion (NO^-) (STAMLER et al. 1992c; BYLER et al. 1983; PARK 1988). The heterolytic mechanism involved can be rationalized by the existence of a thionitrite resonance form in which the NO group accomodates the negative charge; by "activating" the derivatized sulfur atom, this pathway may also serve to potentiate disulfide formation

$$RS-N=O \underset{H^+}{\overset{H^+}{\rightleftarrows}} [RS^+ = N–O^- (H^+)] \xrightarrow{RS^-} RSSR + HNO.$$

The long forgotten report of hydroxylamine isolation from RS-NO (TASKER and JONES 1909) provides further support for this mechanism, as it is a major product of the reaction of singlet NO^- and thiol (TURK and HOLLOCHER 1992)

$$HNO + 2RSH \rightarrow RSSR + NH_2OH.$$

What are the biological implications of this isomerism (RS-N=O $\rightleftarrows$ RS$^+$=N–O$^-$)? In light of many vicinal SH groups essential for protein activity, the formation of RS-NO suggests a mechanism by which "nitric oxide" may regulate protein function by potentiating disulfide formation. Moreover, the propensity of RS-NO for transnitrosation reactions and disulfide formation, taken together with the strong thiol reducing environment in cells, suggests that heterolytic cleavage mechanisms may contribute significantly to RS-NO metabolism in biological systems (STAMLER et al.1992a; SIMON et al. 1993; BUTLER and ASKEW 1993; STAMLER 1994). Thionitrites are also susceptible to oxidation, the latter process leading to formation of thionitrates (RS-NO$_2$) (OAE and SHINHAMA 1983). Nevertheless, the ease by which thionitrites are reduced suggests that reductive pathways are more likely in vivo.

3 Biological Actions of *S*-Nitrosothiols

The inhibition of bacterial growth is an essential function of nitrite during the food curing process. Early studies on the mechanism of bacterial inhibition show that an activation of nitrite is required and that thiol supports such activation (CASTELLANI and NIVEN 1955; JOHNSTON and LOYNES 1971). Independent studies of Mirna and Hoffman indicate that thionitrites form in nitrite-cured meats and that such reactions are very dependent on pH (MIRNA and HOFFMAN 1969). These observations led to the hypothesis that RS-NO may account for the bacteriostatic action of nitrite (KANNER 1979). In a series of important studies INCZE and colleagues (1974) and MORRIS and HANSEN established inhibitory effects of RS-NO on *Streptococcus faecium*, *Salmonella* strains, *Clostridium sporogenes* and *Bacillus cereus*, also learning much about the mechanism of action (MORRIS et al. 1984; MORRIS and HANSEN 1981). In particular, the inhibitory actions of RS-NO on spore outgrowth are correlated with the electron withdrawing effects of the thiyl group (MORRIS and HANSEN 1981) and were shown to be independent of RS-NO size, shape, charge, hydrophobicity and membrane permeability. In light of the above discussion on the chemistry of these compounds, these data suggest a mechanism of action independent of spontaneous homolytic cleavage of the RS-NO to yield NO·. In fact, reports by MORRIS and HANSEN firmly establish that the antimicrobial action of RS-NO results from covalent modification of bacterial membrane sulfhydryl groups (MORRIS et al. 1984; MORRIS and HANSEN 1981), and direct inhibition of sulfhydryl-dependent enzymes, such as glyceraldehyde phosphate dehydrogenase, may further contribute to their bacteriostatic activity (O'LEARY and SOLBERG 1976). Moreover, cell impermeable RS-NO potently inhibits growth of the malarial parasite, *Plasmodium falciparum*, whereas a saturated solution of NO˙ does not (ROCKETT et al. 1991). These data support the interpretation of NO⁺ transfer to critical membrane and cellular thiols leading to formation of protein RS-NO (not excluding subsequent evolution to disulfide), thereby disrupting vital microbial functions. Approximately 20 years later, Fung and coworkers would report that the vasodilatory effects of RS-NO could not be correlated with size, shape, charge, hydrophobicity or lipophilicity and were entirely independent of the rate of NO generation (KOWALUK and FUNG 1990); we have since reported the same for airway smooth muscle relaxation (GASTON et al. 1994b). Moreover, we and others have now shown that RS-NOs react with the cell membrane, both with enzymes that are poorly defined (KOWALUK and FUNG 1990; RADOMSKI et al. 1993) and in chemical reactions with surface thiol groups (SIMON et al. 1993; BUTLER and ASKEW 1993; LIPTON et al. 1993; STAMLER 1994). Notwithstanding these cogent observations, there remains a consensus of opinion that the actions of RS-NO occur solely through "spontaneous" liberation of nitric oxide, the effects of which are in turn mediated through activation of cytosolic guanylate cyclase.

Ignarro and coworkers are to be credited with the first demonstrations that RS-NO are likely intermediates in the metabolism of organic nitrites, nitrates and perhaps nitroprusside (IGNARRO and GRUETTER 1980; GRUETTER et al. 1981; IGNARRO et al. 1980, 1981; IGNARRO 1989). Notwithstanding current controversy on the metabolism of organic nitrates, there are several lines of reasoning supporting the role of RS-NO: (a) incubation of nitroglycerin, organic nitrites, and nitroprusside with thiol all lead to formation of thionitrites under physiological conditions (IGNARRO et al. 1981; FEELISCH 1991); (b) thionitrites markedly activate guanylate cyclase and elevate cyclic GMP in tissue (IGNARRO 1989; STAMLER et al. 1992c); (c) thiols potentiate vasodilatory (IGNARRO et al. 1981) and antiplatelet (LOSCALZO 1985; STAMLER et al. 1988; STAMLER and LOSCALZO 1991) effects of organic nitrites and nitroprusside and these effect can be mimicked by infusion of the corresponding RS-NO in animals (SCHARFSTEIN et al. 1993a; KEANEY et al. 1993); (d) RS-NOs are potent relaxants of vascular smooth muscle in vitro and decrease systemic arterial pressure when infused in animal models; further, they exhibit hemodynamic profiles that closely resemble those of nitroglycerin and nitroprusside (IGNARRO et al. 1981; KEANEY et al. 1993), and finally, (e) activation of purified guanylate cyclase by nitroglycerin, nitroprusside, and nitrite shows a critical dependence on the presence of thiol, under conditions promoting the formation of RS-NO (FEELISCH and NOACK 1987; IGNARRO 1989; STAMLER and LOSCALZO 1991). These findings certainly suggest that RS-NOs, at the very least, can act as intermediates in the activation of guanylate cyclase. A more direct role is certainly not excluded.

Interestingly, very few studies have actually examined the reaction between organic nitrates and sulfhydryl compounds. In early work, IGNARRO reported on the formation of *S*-nitroso-cysteine, when nitroglycerin was incubated with excess cysteine for prolonged periods (IGNARRO and GRUETTER 1980; IGNARRO et al. 1981). The mechanism involved was not studied, but formation of RS-NO occurs only with cysteine and *N*-acetylcysteine and not with other thiols (IGNARRO and GRUETTER 1980; FEELISCH and NOACK 1987). The most plausible first step in this reaction involves nucleophilic attacks by RS⁻ (thiolate) on the positively charged *N*-atom of the nitrate leading to formation of a thionitrate (RS-NO₂) (YEATES et al. 1985; LIPTON et al. 1993). This intermediate would then be subject to further reduction yielding a thionitrite. An interesting possibility might involve isomerization of the thionitrate to form a sulfonylnitrite (RSONO) through migration of the oxygen atom from nitrogen to sulfur. This species would then be susceptible to further nucleophilic attack by thiol yielding RS-NO; nitration of RSH by OONO⁻ may lead to RS-NO by the same pathway (WU et al. 1994; MOHR et al. 1994). Notwithstanding these possibilities, other mechanisms of reduction may occur in cells or in plasma to explain the thiol-mediated catalysis of RS-NO formation reported in biological systems (FUNG et al. 1988; GASTON et al. 1993). The mechanism of RS-NO formation from nitroprusside and organic nitrites (IGNARRO and GRUETTER 1980; FEELISCH and NOACK 1987) is easier to reconcile and likely involves direct nucleophilic attacks on the positively charged nitrogen of the NO group present in both compounds.

It is the general consensus of opinion that nitroso-vasodilators mediate their vascular smooth muscle relaxant effects by activation of cytosolic guanylate cyclase (IGNARRO 1989). The mechanism of activation of guanylate cyclase has been rationalized through formation of the common active species nitric oxide, which interacts with the heme center of the enzyme in the pathway of activation (IGNARRO 1989). In actuality, the molecular details of this mechanism are scanty and poorly understood. In particular, the assumption that RS-NO undergoes homolytic cleavage yielding NO· to activate the enzyme (IGNARRO and GRUETTER 1980) has been invalidated in certain experiments (KOWALUK and FUNG 1990; WILLIAMS 1988). Moreover, there is an extensive literature on the importance of enzyme thiol groups for its activity. The oxidation of protein thiols can lead directly to activation of guanylate cyclase, and, in concert with the reaction of the NO group at the heme center, leads to maximal enzyme activation (STAMLER et al. 1992c) NIROOMAND et al. 1989; KAMISAKI et al. 1986). Further, studies with nitroprusside reveal that activation of guanylate cyclase with nitroprusside promotes mixed disulfide formation involving thiol groups at the enzyme active site (NIROOMAND et al. 1989; KAMISAKI et al. 1986). Taken together, these observations suggest that S-nitrosation of guanylate cyclase may serve a role in activating the enzyme by facilitating the formation of disulfide. While this hypothesis remains to be tested, it is difficult to imagine—under most assay conditions in which thiol is included at millimolar concentrations—that NO group transfer to functional groups distinct from the heme iron would not be taking place.

The pharmacological effects of RS-NOs have been described in other tissues. *S*-nitroso-glutathione, *S*-nitroso-cysteine, *S*-nitroso-homocysteine, *S*-nitroso-penicillamine and *S*-nitroso-albumin are all potent relaxants of both guinea pig and human airway smooth muscle (JANSEN et al. 1991; GASTON et al. 1994b). As reported in vascular tissue, potency in no way correlates with lipophilicity, size or rate of nitric oxide generation (GASTON et al. 1994b). The role of guanylate cyclase in the mechanism of airway relaxation is less clear than in vascular smooth muscle (GASTON et al. 1994a,b). RS-NOs activate airway guanylate cyclase; however, the lack of effect of guanylate cyclase inhibitors in preventing RS-NO relaxation and temporal discrepancies between the rise in cyclic GMP and relaxation responses have resulted in much controversy in this area. RS-NOs have also been shown to relax smooth muscle in the gastrointestinal tract (SLIVKA et al. 1994) and corpus cavernosum of the penis (Saenz de Tejada, personal communication). In the last regard, there is evidence implicating RS-NO as the bovine retractor penis inhibitory factor (KERR et al. 1992).

4 S-Nitrosothiols in Control of Protein Function

In the early studies described above, the reaction of NO with protein thiol groups was envisioned as an adverse consequence of food curing with nitrite (MASSEY et al. 1978; O'LEARY and SOLBERG 1976). Renewed interest in the reaction of NO_x with the SH groups of proteins has been stimulated by the discovery of endogenous nitrogen oxide production in mammalian cells. We have recently focused on the possibility that posttranslational modification of proteins by NO, through reaction with SH groups, may regulate their function (STAMLER et al. 1992a,c,d,e; LIPTON et al. 1993). In a series of studies we demonstrated that proteins of diverse nature and function can be S-nitrosylated, that the derivatized proteins are quite stable at neutral pH, and that this modification confers on proteins smooth muscle (vascular and airway) relaxant and antiplatelet bioactivity (STAMLER et al. 1992a,d,e; SIMON et al. 1993; GASTON et al. 1993, 1994b). S-nitroso-albumin was subsequently identified in plasma as a reservoir of NO and is envisioned to serve a buffer-like function regulating levels of free nitric oxide for the determination of vasomotor tone (STAMLER et al. 1992b). When given by intravenous administration, S-nitroso-albumin markedly reduces mean arterial pressure and inhibits ex vivo platelet aggregation, supporting its potential biological role in modulating vasomotor tone (KEANEY et al. 1993). By comparison with nitroglycerin and nitroprusside, S-nitroso-albumin exhibits prolonged vasodilatory effetcs consistent with the notable stability of S-nitrosoproteins in general (KEANEY et al. 1993). More recently, S-nitrosoproteins were identified in human airway lining fluid in disease states associated with impairment of alveolar-capillary membrane integrity (GASTON et al. 1993). This finding is consistent with the presence of albumin and cysteine-rich mucus proteins in the airways. Although their biological role remains to be ascertained, RS-NO possesses bronchodilator activity and may limit the putative toxicity of NO˙ in the O_2-rich environment of the lung (GASTON et al. 1993).

The idea that protein function may be regulated by NO has early support from several groups of investigators. As detailed above, MORRIS and HANSEN (1981) showed that the bacteriostatic effects of nitrogen oxides resulted from S-nitrosation of protein membrane thiols. In our earliest experiments we directly studied the effects of S-nitrosylation on the activity of two enzymes: tissue-type plasminogen activator (tPA) an endothelium derived fibrinolytic agent with a single free thiol in a domain distant from the active site, and cathepsin B, a sulfhydryl protease in which the thiol is critical for enzyme activity (STAMLER et al. 1992a,d,e). In the case of tPA (STAMLER et al. 1992d), S-nitrosylation enhanced enzyme activity, albeit slightly, while conferring vasodilatory and antiplatelet activity on the native protein. Most importantly, the fibrin specificity of tPA is preserved, providing a means by which NO can be selectively delivered to a fibrin clot, at which site a deficit of NO would be predicted because of endothelial damage. In contrast, the enzymatic activity of cathepsin B is inhibited by S-nitrosylation of active site thiol (STAMLER et al. 1992e). Similar NO-dependent inhibitory effects have been observed with other enzyme. These include aldolase (O'LEARY and SOLBERG 1976), alcohol dehydrogenase (PARK 1988), γ-glutamylcysteinyl synthetase (HAN et al. 1994) and glyceraldehyde-3-phosphate dehydrogenase (GAPDH) (O'LEARY and SOLBERG 1976; MOLINA Y VEDIA et al. 1992; MOHR et al. 1994), the latter event also associated with an ADP-ribosylation-like modification (MOLINA Y VEDIA et al. 1992; STAMLER 1994) in vivo. Studies on the mechanism of GAPDH modification reveal that S-nitrosylation of an active site thiol is requisite for subsequent modification by nicotine adenine dinucleotide phosphate (NAD$^+$) (MOHR et al. 1994). Thus, the formation of protein RS-NO serves a catalytic function. Interestingly, the S-NO intermediate of GAPDH formed during labelling with NAD is devoid of enzymatic activity (MOLINA Y VEDIA et al. 1992), analogous to cathepsin B. That only a fraction of S-nitrosylated protein undergoes covalent modification, taken together with the reversible nature of the S-NO modification, suggests that enzyme activity is regulated by S-nitrosylation (MOLINA Y VEDIA et al. 1992; MCDONALD and MOSS 1993; STAMLER et al. 1992a,c); nonenzymatic NAD-dependent modification of cysteine residues may represent a biological marker for intracellular protein S-nitrosylation. Intriguingly, other proteins such as actin and albumin undergo similar covalent modification of active site thiol, on treatment with NO (CLANCY et al. 1993; DIMMLER and BRUNE 1992).

These ideas have been recently extended to a cell system in which the thiol groups under study comprise the "redox modulatory" site of the N-methyl-D-asparte (NMDA)-type glutamate receptor and determine its ligand-gated activity (LIPTON et al. 1993). Conditions supporting S-nitrosylation of receptor thiol were shown to decrease the calcium influx induced by NMDA binding and the consequent toxicity to surrounding cells. It is not clear from these studies whether S-nitrosylation is sufficient to modulate the activity in the system or whether subsequent oxidation of protein RS-NO to disulfide is required. However, the finding that the reversal of inhibition of NMDA occurs in a two-phase process (LEI et al. 1992) suggests that both direct modification by NO (rapidly reversible) and the promotion of disulfide formation (long-lasting inhibition) occur to modulate the response. Analogous activation of a Ca$^+$ dependent potassium channel on smooth muscle that facilitates relaxation was recently reported (BOLOTINA 1994). That S-nitrosylation of critical thiols on the receptor produced an effect discordant from alkylation suggests that the molecular mechanism of activation involves disulfide formation. Specifically, only the NO group can accommodate a negative charge, and thereby promote the activation-associated oxidation that leads to a conformational switch. The presence of critical thiol groups on other proteins, enzymes—including guanylate cyclase—and surface receptors suggests a common mechanism by which NO can modulate protein function that also shares analogy with the more classical cellular signal transduction pathways. Thus, for example, S-nitrosylation of proteins may act as a rapid reversible signal transduction mechanism, akin, perhaps, to phosphorylation. If true, NO group transfer reactions may be enzymatically controlled in addition to the chemically controlled processes described here.

5 *S*-Nitrosothiols and NO˙ Mediated Toxicity

There is a great deal of concern over the potential toxicity of nitric oxide. Several mechanisms have been described by which NO may exert its cytotoxic effects. Nitric oxide may participate in nitrosative reactions leading to deamination of DNA and mutagenesis (WINK et al. 1991), and reactions of nitric oxide with superoxide lead to the formation of the highly toxic oxidant peroxynitrite (BECKMAN et al. 1990). In addition, the direct reaction of nitric oxide with iron sulfur centers in proteins may lead to the inhibition of their activity (HIBBS 1991). These concerns are supported by data showing that nitric oxide exposure can cause severe lung injury (WORLD HEALTH ORGANIZATION 1977; GASTON et al. 1994a), promote the formation of tumors (WORLD HEALTH ORGANIZATION 1977), and enhance neuronal destruction in stroke and other neurodegenerative disorders (SNYDER 1993). The reaction of nitrogen oxides (NO_x) with thiols may provide insights into divergent protective responses. Certainly, one of the main functions of biological thiols is to detoxify oxygen-derived free radicals and peroxides. By extension, thiols may play an important protective role in disarming nitric oxide. Levels of glutathione rise in lung tissue in response to oxidant stress, such as exposure to high levels of NO contained in cigarette smoke, and endogenous RS-NO levels increase in the lung in response to immune activation and inflammation (GASTON et al. 1993, 1994a). Moreover, the relative stability of these endogenous RS-NO (on the order of hours) may serve to limit the reactions of NO˙ with oxygen and superoxide that are associated with nitric oxide mediated toxicity (GASTON et al. 1993, 1994a; STAMLER et al. 1992c). The added finding that RS-NO possesses bronchodilator activity suggests a means by which the NO group is packaged in biological systems to preserve its bioactivity (GASTON et al. 1993, 1994b). Recent work indicates that NO can inhibit glutathione synthesis by inactivation of γ-glutamylcysteinyl synthetase (HAN et al. 1994), which catalyses the initial and rate-limiting step. Thus, the potential for regulation of cellular functions influenced by redox status notwithstanding, excessive NO may facilitate oxidative injury by depleting the thiol pool.

In the central nervous system, similar protective effects of RS-NO have been observed. Nitric oxide mediated toxicity has been attributed to the formation of peroxynitrite by way of reaction with superoxide (LIPTON et al. 1993). In contrast, formation of RS-NO provides a mechanism for limiting the reaction of NO· with O_2^-. As discussed above, S-nitrosation reactions may have additional protective effects by down-regulating NMDA receptor activity, which is a prerequisite for nitric oxide synthesis (LIPTON et al. 1993; SNYDER 1993). These findings suggest that, with excessive generation of nitric oxide by neurons, the molecule may feedback to inhibit its own production (MANZONI et al. 1992).

Paradoxically, biological thiols may exert toxic effects as well. Homocysteine, for example, will readily reduce oxygen, resulting in formation of superoxide and hydrogen peroxide (STAMLER et al. 1993; STARKEBAUM and HARLAN 1986); metal ions such as Cu^{2+} and Fe^{2+} may serve as catalysts in these reactions, possibly forming thiol-metal complexes (JOCELYN 1972)

$$Fe(II)\ (SR)_2 + O_2 \rightarrow Fe(III)\ (SR)_2 + O_2^-,$$

$$O_2^- + RS^- \rightarrow O_2^= + RS^·,$$

$$Fe(III)\ (SR)_2 + RS^- \rightarrow Fe(II)\ (SR)_2 + RS^·,$$

$$2\ RS^· \rightarrow RSSR$$

Reduced oxygen species are highly toxic to endothelial cells and are believed to play an important role in the vascular related morbidity seen in patients with hyperhomocysteinemia (STAMLER et al. 1993; STARKEBAUM and HARLAN 1986), an independent risk factor for atherosclerosis. Under normal physiological circumstances the interaction between endothelium-derived relaxing factor (EDRF) and homocysteine results in the formation of S-nitroso-homocysteine, a potent vasodilator and antiplatelet agent (STAMLER et al. 1993). With the thiol group "blocked" by NO, homocysteine is also prevented from generating hydrogen peroxide (STAMLER et al. 1993). However, with elevated levels of homocysteine incurred through an acquired or inborn alteration of metabolism, the cytotoxic properties of the molecule may embarrass EDRF production, setting a cycle in motion in which the antithrombotic, cytoprotective mechanism of S-nitrosation is increasingly compromised. Thus, S-nitrosation may represent a protective cell regulatory mechanism which simultaneously confers upon biological thiols EDRF-like bioactivity.

6 S-Nitrosothiols In Vivo

S-nitrosothiols have been identified in biological systems by several groups of investigators. Fung and coworkers identified low molecular weight RS-NOs in the plasma of human subjects treated with nitroglycerin (FUNG et al. 1988). These experiments also showed that metabolism of nitroglycerin was dependent, in part, on the thiol group of albumin, suggesting that protein RS-NOs may also form (FUNG et al. 1988; CHONG and FUNG 1990). We have extended these findings by showing that S-nitroso-proteins are present in the plasma of normal subjects (STAMLER et al. 1992b) and in the exudate of airway lining fluid in selected disease states (GASTON et al. 1993). In normal airways, S-nitroso-glutathione has been identified in concentrations sufficient to influence basal airway tone (GASTON et al. 1993). S-nitroso-glutathione also forms intracellularly in neutrophils on exposure to NO gas and with cell activation (CLANCY et al. 1994). Interestingly, Ribeiro and colleagues identified nitrosovasodilator activity in the salivary gland of *Rhodnius prolixus*, of which approximately 10%–20% was attributed to S-nitrosothiol (RIBEIRO et al. 1993). Finally, it has been estimated that a significant percentage of NO_x in the gastrointestinal tract exists in the form of RS-NO (MIRNA and HOFMANN 1969; KANNER 1979; MASSEY et al. 1978).

The hemodynamic effects of RS-NOs have received recent attention. In early reports, Ignarro's group showed vasodilator effects in the cat (IGNARRO et al. 1981). S-nitroso-N-acetylpenicillamine, S-nitroso-cysteine, S-nitroso-mercapto-ethylamine and S-nitroso-3-mercaptopropionic acid each decreased systemic arterial pressure in a dose-related fashion (IGNARRO et al. 1981). The onset of action was immediate in each case and the reported duration of action on the order of approximately 1–3 min. S-nitroso-cysteine was the most potent of the RS-NOs with a dose-response curve also lying to the left of that of nitroglycerin and nitroprusside. The potency of S-nitroso-cysteine has been confirmed by KEANEY and colleagues (1993). S-nitroso-cysteine caused marked changes in blood pressure when given by intravenous bolus infusion and very dramatic changes in coronary flow, indicative of its vasodilator effects on microvessels. In the same study, the authors investigated the effects of S-nitroso-albumin, which also exhibited vasodilator and antiplatelet properties reminiscent of ERDF. However, the kinetic profile and hemodynamic response was unique. S-nitroso-albumin had a greater duration of action than either nitroglycerin or S-nitroso-cysteine when given by intravenous infusion and a much longer duration of action on coronary blood flow when infused directly into the coronary artery. It appears that the actions of S-nitroso-albumin are mediated by way of NO groups transfer to more reactive, low-molecular weight thiol pools in plasma (STAMLER et al. 1992b; SCHARFSTEIN et al. 1993b), which in turn, may target the NO group to the vascular (or platelet) surface (FUNG et al. 1988; SIMON et al 1993; RADOMSKI et al. 1993). SCHAFER and coworkers (1991) studied the hemodynamic effects of S-nitroso-captopril in the anesthetized dog. Interestingly, the nitrovasodilator effect was ten- to 30-fold less potent than nitroglycerin when administered by bolus, but significantly more effective than the former when given by continuous infusion. The duration of action of S-nitroso-captopril was also significantly greater than that of nitroglycerin, which for unclear reasons, significantly exceeded the half-life of the drug. The attenuated pressure response to angiotensin in these animals indicates that this agent also manifests angiotensin converting enzyme inhibitor activity, which contributes to its mechanism of action. Recently, Vallance and coworkers infused S-nitroso-glutathione into the forearm of humans (Vallance, personal communication). The antiplatelet effects of the drug was significantly more impressive than its vasodilator response implicating different pathways of metabolism in the platelet and vascular wall.

The effects of RS-NOs on human gastrointestinal motility (SLIVKA et al. 1993) and airway tone (GASTON et al. 1993, 1994a, 1994b) have also been recently studied. When applied topically to the duodenum and colon, S-nitroso-N-acetylcysteine causes profound inhibition of intestinal motility (SLIVKA et al. 1994). Application directly to the sphincter of Oddi lowers sphincter of Oddi pressure and inhibits the intrinsic frequency of contraction. At the concentration of 100 μM applied in 10 ml of saline, the onset of action occurs in minutes and the duration of action was estimated on the order of 10–20 min. Studies examining the bronchodilator action of S-nitroso-N-acetylcysteine are currently underway. As in the gastrointestinal tract, topical administration (nebulized in saline) was not associated with systemic side effects. Early results in subjects with mild asthma should become available in the near future. RS-NOs have also been used on ureteral smooth muscle and corpus cavernosum, suggesting their potential uses in the treatment of ureteral spasm and erectile dysfunction (Saenz de Tejeda, personal communication).

7 Summary

The reactivity of selected RS-NOs has led to the misconception that these compounds are uniformly unstable under physiological conditions. Moreover, current evidence supports the notion that biological responses elicited by RS-NOs may result from either liberation of nitric oxide or from NO group transfer chemistry involving either NO^+ or NO^-. Some evidence suggests that such reactions may be enzymatically controlled. The data supporting the potential biological relevance of RS-NOs include : (1) evidence that these compounds form under physiological conditions; (2) their identification in insects, lower mammals, and several human biological systems; and (3) findings that RS-NOs possess a wide range of biological activities, including antimicrobial effects, vasodilation, platelet inhibition, bronchodilation and inhibition of intestinal motility, while being relatively resistant to reactions with O_2 and O_2^- associated with NO˙ toxicity. It is further noteworthy that biological activity of RS-NO is often not related to the propensity to liberate NO˙, and these adducts are generally more potent and selective in their action than NO˙ itself (STAMLER et al. 1989; COOKE et al. 1990; ROCKETT et al. 1991; JANSEN et al. 1991; LIPTON et al. 1993).

The data presented here support the idea that RS-NO may be involved in stabilizing nitric oxide-like bioactivity, in transporting and targeting the NO group to specific (thioregulatory) effector sites, in mitigating the cytotoxic effects of nitric oxide that result from reaction with oxygen species, and may serve to regulate protein function in a posttranslational modification akin, perhaps, to phosphorylation. The recently demonstrated NO group transfer reactions to plasma membrane proteins containing reactive sulfhydryls (Lɪᴘᴛᴏɴ et al. 1993; Sᴛᴀᴍʟᴇʀ 1994) also raises the possibility of signal transduction initiated through more traditional "agonist-receptor" mediated pathways.

Acknowledgements. I am indebted to Professor David J. Singel for his critical review of this manuscript and for related discussions. JSS is a Pew Scholar in the Biomedical Sciences and the recipient of a Clinical Investigator. Award from the National Institutes of Health (HL 02582-011)

References

Al-Kaabi SS, Williams DLH, Bonnet R, Ooi S (1982) A kinetic investigation of the thionitrite from (±)-2-acetylamino-2-carboxy-1,1-dimethylethanethiol as a possible nitrosating agent. J Chem Soc Perkins II: 227–230

Aldred SE, Williams LH (1982) Kinetics and mechanisms of the nitrosation of alcohols, carbohydrates, and a thiol. J Chem Soc Perkin Trans II: 777–782

Beckman JS, Beckman TW, Chen J, Marshell PA, Freeman BA (1990) Apparent hydroxyl radical production by peroxynitrite: implications for endothelial injury from nitric oxide and superoxide. Proc natl Acad Sci USA 87: 1620–1624

Bolotina VM, Najibi S, palacio JJ, Pagano PJ, Cohen RA (1994) Nitric oxide directly activates calcium-dependent potassium channels in vascular smooth muscle. Nature 368: 850–853

Bonnett R, Holleyhead R, Johnson BI, Randall EW (1975) Reactions of acidified nitrite solutions with peptide derivatives: evidence for nitrosamine and thionitrite formation from ^{15}N N.M.R. studies. J Chem Soc Perkin I: 2261–2264

Butler AR, Askew SC (1993) The vascular action of S-nitroglutathione: evidence for NO transfer. Endothelium 1: 144A

Byler DM, Gosser DK, Susi H (1983) Spectroscopic estimation of the extent of S-nitrosothiol formation by nitrite action on sulfhydryl groups. J Agric Food Chem 31: 523–527

Castellani AG, Niven CF (1985) Factors affecting the bacteriostatic action of sodium nitrite. Appl Microbiol 3: 154–159

Clancy RM, Piziak-Leszcynska J, Abramson SB (1993) Nitric oxide stimulates ADP-ribosylation of action in human neutrophils. Biochem Biophys Res Commun 191: 847–852

Clancy RM, Yegudin J, Levartovsky D, Piziak-Leszcynska J, Abramson SB (1994) Nitric oxide reacts with intracellular glutathione and activates the hexose monophosphate shunt in human neutrophils: evidence for S-nitrosoglutathione as a bioactive intermediary. Pro Natl Acad Sci USA 91: 3680–3684

Chong S, Fung HI (1990) Thiol-mediated catalysis of nitroglycerin degradation by serum proteins. Drug Metab Dispos 18: 61–67

Chong S, Fung H (1991) Biochemical and pharmacological interactions between nitroglycerin and thiols. Biochem Pharmacol 42: 1433–1439

Cook JP, Stamler JS, Andon N, Davies PF, McKinley G, Loscalzo J (1990) Flow Stimulation endothelial cells to release a nitrosovasodilator that is potentiated by reduced thiol. Am J Physiol 28: H804–H812

Dimmler S, Brune B (1992) Characterization of a nitric oxide catalysed ADP ribosylation of glyceraldehyde-3-phosphate dehydrogenase. Eur J Biochem 1202: 305–310

Feelisch M (1991) The biochemical pathways of nitric oxide formation from nitrosovasodilators; Appropriate choice of exogenous NO donors and aspects of preparation and handling of aqueous NO solutions. J Cardiovasc Pharmacol 17 Suppl 3: S25–S33

Feelisch M, Noack E (1987) Nitric oxide formation from nitrovasodilators occurs independently of hemoglobin or non-heme iron. Eur J Pharmacol 142: 465–469

Field L. Dilts RV, Ravichandran R, Lenhert PG, Carnahan GE (1978) An unusually stable thionitrite from N-acetyl-D, L-penicillamine; X-ray crystal and molecular structure of 2-(acetylamino)-2-carboxy-1, 1-dimethylethyl thionitrite. J Chem Soc Chem Comm: 249–250

Fung HL, Chong S, Kowaluk E, Hough K, Kakemi M (1988) Mechanisms for the pharmacologic interaction of organic nitrates with thiols. Existence of an extracellular pathway for the reversal of nitrate vascular tolerance by N-acetylcysteine. J Pharmacol Exp Ther 245: 524–530

Gaston B, Reilly J, Drazen JM, Fackler J, Ramdev P, Arnelle D, Mullins M, Sugarbaker DJ, Chee C, Singel DJ. Loscalzo J, Stamler JS (1993) Endogenous nitrogen oxides and bronchodilator S-nitrosothiols in human airways. Proc Natl Acad Sci USA 90: 10957–10961

Gaston B, Drazen JM, Loscalzo J, Stamler JS (1994a) The biology of nitrogen oxides in the airways. State-of-the-Art. Am Rev Respir Dis 149: 538–551

Gaston B, Drazen JM, Jansen A, Sugarbaker DJ, Loscalzo J, Richards W, Stamler JS (1994b) Relaxation of human bronchial smooth muscle by S-nitrosothiols in vitro. J Pharmacol Exp Ther 268: 978–984

Gruetter CA, Gruetter DY, Lyon JE, Kadowitz PJ, Ignarro LJ (1981) Relationship between cyclic guanosine 3':5'-monophosphate formation and relaxation of coronary arterial smooth muscle by glycerol trinitrate, nitroprusside, nitrite and nitric oxide: effects of methylene blue and methemoglobin. J Pharmacol Exp Ther 219: 181–186

Han J, Stamler JS, Griffith O (1994) Inhibition of γ-glutamylcysteine synthetase by nitirc oxide donors. FASEB J 8: A1288

Hibbs JB (1991) Overview of cytotoxic mechanisms and defense of the intracellular environment against microbes. The biology of Nitric oxide II. Portland, Chapel Hills, pp 201–206

Incze K, Parkes J, Mihalyi V, Zukal E (1974) Antibacterial effect of cysteine-nitrosothiol and possible precursors thereof. Appl Microbiol 27: 202–205

Ignarro LJ (1989) Biological actions and properties of endothelium-derived nitric oxide formed and released from artery and vein. Circ Res 65:1–21

Ignarro L, Gruetter CA (1980) Requirement of thiols for activation of coronary arterial guanylate cyclase by glycerol trinitrate and sodium nitrite. Biochim Biophys Acta 631: 221–231

Ignarro LJ, Edwards JC, Gruetter DY, Barry BK, Gruetter CA (1980) Possible involvement of S-nitrosothiols in the activation of guanylate cyclase by nitroso compounds. FEBS Lett 110: 275–278

Iganarro LJ, Lippton H, Edwards JC, Baricos WH, Hyman AL, Kadowitz PH, Gruetter CA (1981) Mechanism of vascular smooth muscle relaxation by organic nitrates, nitrites, nitroprusside and nitric oxide: Evidence for the involvement of S-nitrosothiols as active intermediates. J Pharmacol Exp Ther 218: 739–749

Jansen A, Drazen J, Osborne JA, Brown R, Loscalzo J, Stamler JS (1992) The relaxant properties in guinea pig airways of S-nitrosothiols. J Pharmacol Exp Ther 261: 154–160

Jocelyn PC (1972) In: Biochemistry of the SH group. Academic, London

Johnson MA, Loynes JS (1971) Inhibition of Clostridium botulinum by sodium nitrite in a bacteriologic medium and in meat. Can Inst Food Technol J 4: 179–184

Josephy PD, Rehorek D, Janzen EG (1984) Electron spin resonance spin trapping of thiyl radicals from the decomposition of thionitrites.Tetrahedron Lett 25: 1685–1688

Kamisaki Y, Waldman SA, Murad F (1986) The involvement of catalytic site thiol groups in the activation of soluble guanylate cyclase by sodium nitroprusside. Arch Biochem Biophys 251:709–714

Kanner J (1979) S-nitrosocysteine (RSNO), and effective antioxidant in cured meat. J Am Oil Chem Soc 56: 74–76

Keaney JF, Simon DI, Stamler JS, Jaraki O, Scharfstein J, Vita JA, Loscalzo J (1993) NO forms an adduct with serum albumin that has endothelium-derived relaxing factro-like properties. J Clin Invest 91: 1582–1589

Kerr SW, Buchanan LV, Bunting S, Mathews WR (1992) Evidence that S-nitrosothiols are responsible for the smooth muscle relaxing activity of the bovine retractor penis inhibitory factor. J Pharmacol Exp Ther 263:285–263

Kowaluk EA, Fung HL (1990) Spontaneous liberation of nitric oxide cannot account for in vitro vascular relaxation by S-nitrosothiols. J Pharmacol Exp Ther 255: 1256–1264

Lecher H, Siefken W (1926) Nitrosyl-derivate des zweiwertigen Schwefels, I: Das Nitrosylethyl-mercaptid. Ber Dtsch Chem Ges 59B: 1314–1326

Lei SZ, Pan ZH, Aggarwal SK, Chen HSV, Hartman J, Sucher NJ, Lipton SA (1992) Effect of nitric oxide production on the redox modulatory site of the NMDA receptor-channel complex. Neuron 8: 1087–1099

Lipton SA, Choi YB, Pan ZH, Lei SZ, Vincent Chen HS, Sucher NJ, Loscalzo J, Singel DJ, Stamler JS (1993) A redox-based mechanism for the neuroprotective and neurodestructive effects of nitric oxide and related nitroso-compounds. Nature 364: 626–632

Loscalzo J (1985) N-acetylcysteine potentiates inhibition of platelet aggregation by nitroglycerin. J Clin Invest 76: 703–708

Manzoni O, Prezeau L, Marin P, Deshager S, Bockaert J, Fagni L (1992) Nitric oxide-induced blockade of NMDA receptors. Neuron 8: 653–662

Mason J (1969) Trithioromethyl thionitrite. J Chem Soc A 1587–1592

Massey RC, Crews C, Davies R, McWeeny DJ (1978) A study of the competitive nitrosations of pyrrolidine, ascorbic acid, cysteine and p-Cresol in a protein-based model system. J Sci Food Agric 29: 815–821

McNainly J, Williams DLH (1993) Fate of nitric oxide from the decomposition of S-nitrosothiols. Endothelium 1: 141A

Mirna A, Hofmann K (1969) Uber den verbleib von Nitrit in fleischwaren. 1. Umsetzung von Nitrit mit sulhydryl verbindungen. Fleischwirtschaft 10: 1361–1364

Molina Y Vedia L, Mcdonald B, Reep B, Brune B, DiSilvio M, Billiar TR, Lapentina EG (1992) Nitric-oxide-induced S-nitrosylation of glyceraldehyde-3-phosphate dehydrogenase inhibits enzymatic activity and increases endogenous ADP ribosylation. J Biol Chem. 267: 24929–24932

Mohr S, Stamler JS, Brune B (1994) Mechanism of covalent modification of glyceraldehyde-3-phosphate dehydrogenase at its active site thiol by nitric oxide, peroxynitrite and related nitrosating agents. FEBS Letters 348: 223–227

Morris SL, Hansen JN (1981) Inhibition of Bacillus cereus spore outgrowth by covalent modifications of a sulfhydryl group by nitrosothiol and iodoacetate. J Bacteriol 148: 465–471

Morris SL, Walsh RC, Hansen JN (1984) Identification and characterization of some bacterial membrane sulfhydryl groups which are targets of bacteriostatic and antibiotic action. J Biol Chem 259: 13590–13594

Myers PR, Minor RL, Guerra R, Bates JN, Harrison DG (1990) Vasorelaxant properties of the endothelium-derived relaxing factor more closely resemble S-nitrosocysteine than nitric oxide. Nature 345: 161–163

Niroomand F, Rossle R, Mulsch A, Bohme A (1989) Under anaerobic conditions, soluble guanylate cyclase is specifically stimulated by glutathione. Biochem Biophys Res Commun 161: 75–80.

Oae S, Shinhama K (1983) Organic thionitrites and related substances. In: Organic preparations and procedures. Organic Prep Proced 15: 165–198

Oae S, Fukushima D, Kim YH (1977) Novel method of activating thiols by their conversion into thionitries with dinitrogen tetroxide. J Chem Soc Chem Comm: 407–408

Oae S, Kim YH, Fukushima D, Shinhama K (1978) New syntheses of thionitrites and their chemical reactivities. J Chem Soc Perkin 1: 913–917

O'Leary V, Solberg M (1976) Effect of sodium nitrite on inhibition of intracellular thiol groups and on the activity of certain glycolytic enzymes in Clostridium porringers. Appl Environ Microbiol 31: 208–212

Park JW (1988) Reaction of S-nitrosoglutathione with sulfhydryl groups in protein. Biochem Biophys Res Commun 152: 916–920

Pryor WA, Church DF, Govinden CK, Crank G (1982) Oxidation of thiols by nitric oxide and nitrogen dioxide: synthetic utility and toxicological implications. J Org Chem 47: 156–159.

Radomski MW, Rees DD, Durta A, Moncada S (1993) S-nitroso-glutathione inhibits platelet activation in vitro and in vivo. Br J Pharmacol, 107: 745–749

Ramdev P, Loscalzo J, Feelisch M, Stamler JS (1993) Biochemical properties and bioactivity of a physiologic NO reservoir. Circulation 88: 1–522

Rao PM, Copeck JA, Knight AR (1967) Reactions of thiyl radicals II. The photolysis of methyl disulfide vapor. Can J Chem 45: 1369–1374

Ribeiro JM, Hazzard JMH, Nussenzveig RH, Champagne DE, Walker FA (1993) Reversible binding of nitric oxide by a salivary heme protein from a bloodsucking insect. Science 260: 539–541

Ridd J (1978) Diffusion control and pre-association in nitrosation, nitration and halogenation. Adv Phys Organ Chem 16: 1–49

Rinden E, Maricq MM, Grabowski JJ (1989) Gas-phase ion-molecule reactions of the nitric oxide anion. J Am Chem Soc II: 1203–1210

Rockett KA, Auburn MM, Lowden WB, Clark IA (1991) Killing of Plasmodium falciparum in vivo by nitric oxide derivatives. Infect Immun 59: 3280–3283

Schafer JE, Lee F, Thomson S, Han BJ, Cooke JR, Loscalzo J (1991) The hemodynamic effects of S-nitrosocaptopril in anesthetized dogs. J Pharmacol Exp Ther 256: 704–709

Scharfstein JS, Keaney J, Stamler JS, Vita J, Loscalzo (1993a) Low molecular weight thiols transfer nitric oxide from an endogenous plasma reservoir to vascular smooth muscle. Clin Res 41: 232A

Scharfstein JS, Slivka A, Stamler JS, Loscalzo J (1993b) In vivo transfer of nitric oxide from a plasma reservoir to cysteine. Circulation 88: 1–523

Simon DI, Stamler JS, Jaraki O, Keaney J, Osborne JA, Francis SA, Singel DJ, Loscalzo J (1993) Antiplatelet properties of protein S-nitrosothiols derived from nitric oxide and endothelium-derived relaxing factor. Arterioscler Thromb 13: 791–799

Slivka A, Loscalzo J, Chuttani R, Kobzik L, Bredt D, Carr-Locke DL, Stamler JS (1994) Inhibition of sphincter of oddi function by the nitric oxide carrier, S-nitroso-N-actylcysteine in rabbits and humans. J Clin Invest (in press)

Snyder SH (1993) Janus faces of nitric oxide. Nature 364: 577

Stamler JS, Loscalzo J (1991) The antithrombotic effects of organic nitrates. Trends Cardiovasc Med 1: 346–353

Stamler JS (1994) Redox Signaling: Nitrosylation and related target interactions of nitric oxide. Cell 78: 931–936

Stamler JS, Loscalzo J (1992) Capillary electrophoretic detection of thiols and their S-nitrosated derivatives. Anal Chem 64: 779–785

Stamler JS, Cunningham M, Loscalzo J (1988) Reduced thiols and the effect of nitroglycerin on platelet function. Am J Cardiol 62: 377–380

Stamler JS, Mendelsohn M, Amarante P, Davies PF, Cooke JP, Loscalzo J (1989) N-acetylcysteine potentiates platelet inhibition by endothelium derived relaxing factor. Circ Res 65: 789–795

Stamler JS, Simon DI, Osborne JA, Mullins ME, Jaraki O, Michel T, Singel DJ, Loscalzo J (1992a) S-nitrosylation of proteins with nitric oxide: synthesis and characterization of biologically active compounds. Proc Natl Acad Sci USA 89: 444–448

Stamler JS, Jaraki O, Osborne J, Simon DI, Keaney J, Vita J, Singel D, Valeria RC (1992b) Nitric oxide circulates in mammalian plasma primarily as an S-nitroso adduct of serum albumin. Proc Natl Acad Sci USA 89: 7674–7677

Stamler JS, Singel D, Loscalzo J (1992c) Biochemistry of nitric oxide and its redox activated forms. Science 258: 1898–1902

Stamler JS, Simon DI, Jaraki O, Osborne JA, Francis J, Mullins M, Singel D, Loscalzo (1992d) S-nitrosylation of tissue-type plasminogen activator confers vasodilatory and antiplatelet properties on the enzyme. Proc Natl Acad Sci USA 89: 8087–8091

Stamler JS, Simon DI, Osborne JA, Mullins M, Jaraki O, Michel T, Singel D, Loscalzo J (1992e) Exposure of sulfhydryl containing proteins to nitric oxide and endothelium-derived relaxing factor confers novel bioactivity and modulates their intrinsic functional properties. In: Moncada S, Marletta MA, Higgs A, Hibbs JB (eds) Biology of nitric oxide I. Portland Chapel Hill, pp 20–23

Stamler JS, Osborne JA, Jaraki O, Rabbani LE, Mullins M, Singel D, Loscalzo J (1993) Adverse effects of homocysteine are modulated by endothelium-derived relaxing factor and related oxides of nitrogen. J Clin Invest 1: 308–318

Starkebaum G, Harlan JM (1986) Endothelial cell injury due to copper-catalyzed hydrogen peroxide generation from homocysteine. J Clin Invest 77: 1370–1376

Tasker HS, Jones HO (1909) The action of mercaptans on acid chlorides, part II. The acid chlorides of phosphorous, sulfur and nitrogen. J Chem Soc 95: 1910

Turk T, Hollocher TC (1992) Oxidation of dithiothreitol during turnover of nitric oxide reductase: evidence for generation of nitroxyl with the enzyme from paracoccus denitrificans. Biochem Biophys Res Commun 183: 983–988

Williams DHL (1988) Nitrosation. Cambridge University Press, Cambridge

Wink DA, Kasprzak KS, Maragos CM, Elespuru RK, Misra M, Dunams TM, Cebula TA, Koch WH, Andrews AW, Allen S, Keefer LK (1991) DNA deaminating ability and genotoxicity of nitric oxide and its progenitors. Science 254: 1001–1003

Wink DA, Darbyshire JF, Nims RW, Saavedra JE, Ford PC (1993) Reactions of bioregulatory agent nitric oxide in oxygenated aqueous media: determination of the kinetics of oxidation and nitrosation by intermediates generated in the NO/O_2 reaction. Chem Res Toxicol 6: 23–27

World Health Organization (1977) Environmental health criteria 4 oxides of nitrogen. World Health Organization, Geneva

Wu M, Kaminski PM, Fayngersh RP, Groszek LL, Pritchard KA, Hintze TH, Stemerman MB, Wolin MS (1994) Involvement of nitric oxide and nitrosothiols in relaxation of pulmonary arteries to peroxynitrite. Am J Physiol 266: H2108–H2113

Yeates RA, Laufen H, Leitold M (1985) The reaction between organic nitrates and sulfhydryl compounds: a possible model system for the activation of organic nitrates. Mol Pharmacol 28: 555–559

A New Nitric Oxide Scavenger, Imidazolineoxyl N-Oxide Derivative, and Its Effects in Pathophysiology and Microbiology

H. Maeda[1], T. Akaike [1], M. Yoshida [2], K. Sato[1], and Y. Noguchi [1]

1 Introduction

The increased appreciation of nitric oxide (NO) as an important molecule in biology and medicine has been accompanied by an unprecedented interest in inhibitors of NO synthesis and NO scavengers. L-Arginine analogues (Fig. 1) have been used widely as competitive inhibitors of NO synthase (NOS), thus inhibiting the function of NO indirectly. Some of these analogues have even been used in human trials, although there are potential problems with in vivo use. L-Arginine analogues may interfere with the urea cycle and protein synthesis, and long-term suppression of both the constitutive (cNOS) and the induced forms (iNOS) of the enzyme might result in impairment of the neural and circulatory systems. Although L-arginine analogues can suppress the biological effect of endothelium-derived relaxing factor (EDRF) stimulators, such as acetylcholine, proof that NO is an EDRF awaits the demonstration of chemical entrapment of NO at the site of action with simultaneous disappearance of its vasorelaxing activity. Meanwhile NO gas or dilute NO solutions have been shown to relax vascular tone, and the identity of EDRF with NO appears essentially unambiguous.

[1] Department of Microbiology, Kumamoto University School of Medicine, Honjo 2-2-1, Kumamoto 860, Japan
[2] Department of Urology, Kumamoto University School of Medicine, Honjo 2-2-1, Kumamoto 860, Japan

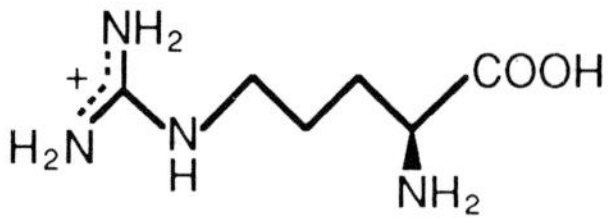

L-Arginine

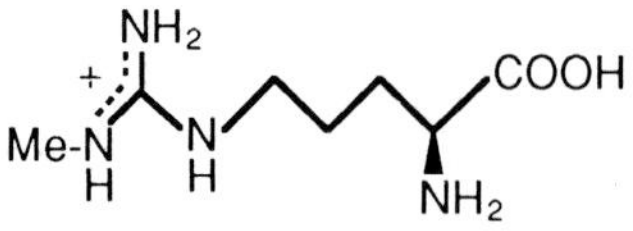

N^{ω}-Monomethyl-L-arginine (L-NMMA)

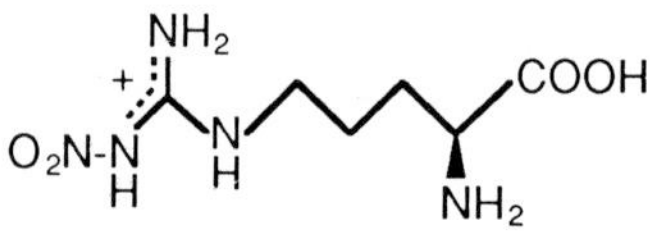

N^{ω}-Nitro-L-arginine (L-NNA)

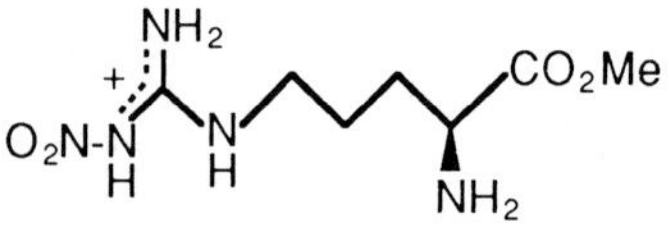

N^{ω}-Nitro-L-arginine methylester (L-NAME)

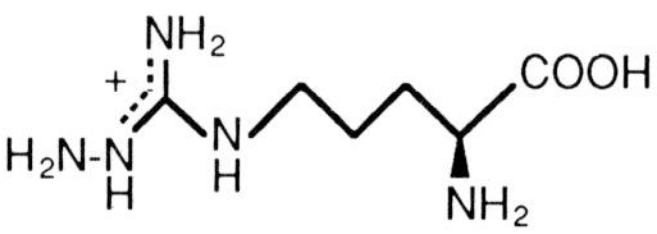

N^{ω}-Amino-L-arginine (L-NAA)

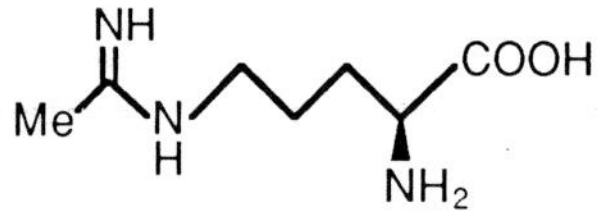

L-N^5-(1-Iminoethyl)-ornithine (L-NIO)

Fig. 1. Structure of L-arginine and various L-arginine analogues, that inhibit nitric oxide synthase

We have recently described an imidazolineoxyl *N*-oxide (PTIO) compound that selectively reacts with NO by converting it to NO_2, while simultaneously and completely abrogating the biological effect of NO. This prompted a detailed investigation of the chemical mechanism by which NO interacts with PTIO, focusing on unique electron spin resonance (ESR) signals, and further functional analysis of NO in vitro and in vivo. We investigated two pathological conditions, endotoxic shock and vascular permeability enhancement in solid tumors, in which NO seems to be involved. Both of these pathological situations are clearly inhibited by PTIO, suggesting the direct causal role of NO and the potential usefulness as a therapeutic strategy based on the administration of PTIO. We have also studied the controversial cytotoxic effect of NO using the bacteria *Staphyloccus aureus* and *Cryptococcus neoformans.*

The biological function of NO has been analyzed primarily through its inhibition of NOS, which has been achieved using glucocorticoids to inhibit NOS induction, (MONCADA et al. 1992), trifluoperazine to inhibit calmodulin-dependent function (for neuronal and endothelial cNOS) (BREDT and SNYDER 1990), and diphenylene iodonium, di-2-thienyliodonium, or iodoniumdiphenyl to inhibit NADPH, an essential electron donor for NOS (STUEHR et al. 1991). However since NADPH is required in many biochemical reactions, such inhibitors have only limited value in an in vivo setting (GATLEY and SHERRATT 1976). Compounds such as 2,4-diamino-6-hydroxyprimidine and N-acetylserotonin, which inhibit biosynthesis of tetrahydrobiopterin, a cofactor of NOS, might selectively inhibit iNOS activity in vivo (GROSS and LEVI 1992; SAKAI et al. 1993), but all of these reagents have potential drawbacks.

Other compounds such as SH- or iron-containing compounds, including glutathione (STAMLER et al. 1992), albumin or hemoglobin (KANNER et al. 1992), might compete with the target molecules of NO or serve as an NO carrier in biological phenomena in which NO is crucially involved. This possibility is discussed in the chapter by Stamler. Here, we focus on PTIO derivatives and functional analyses of NO using PTIOs in vitro, ex vivo, and in vivo.

2 Selective Reaction of Imidazolineoxyl N-Oxide Derivatives with Nitric Oxide

Figure 2 shows the chemical structure of PTIO and the proposed reaction equation, in which R can be replaced with various groups. Synthesis of these PTIO derivatives has been described (AKAIKE et al. 1993). The derivatives 2-phenyl-4,4,5,5-tetramethylimidazoline-1-oxy-3-oxide (PTIO), carboxy-PTIO and carboxymethoxy-PTIO are the most well studied. Figure 3 shows changes in the electron spin resonance (ESR) spectra obtained from PTIO alone before reaction with NO, the purified reaction product, i.e. 2-phenyl-4,4,5,5-tetramethyl-imidazoline-1-oxyl (PTI), and the reaction mixture of PTIO and NO. Computer simulation resolves the latter spectrum into two components similar to the spectra in (a) and (b). Thin-layer chromatography revealed a single spot of PTI after complete reaction of PTIO and NO, with a different mobility from PTIO. The Rf value of the PTI product was 0.48, with a yellow color, whereas Rf of the original PTIO was ~0.30, with a blue color as determined in a n-hexane/chloroform/methanol system (10:5:1, v/v). The amount of PTI formed was equal to that of PTIO consumed, and NO reacted with PTIO stoichiometrically under anaerobic

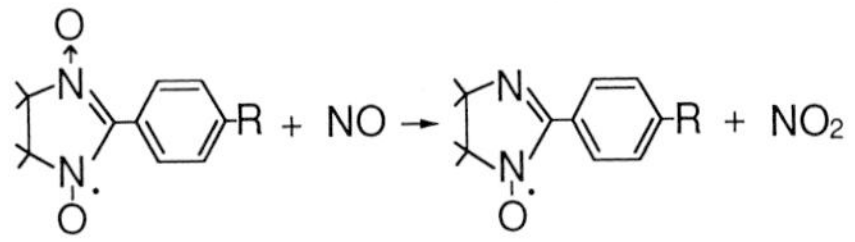

Fig. 2. Structures of imidazolineoxyl N-oxide (PTIO) derivatives (left) and reaction scheme of PTIOs with nitric oxide (NO) (R,H:2-phenyl-4,4,5,5-tetramethyl-imidazoline-1-oxyl-3-oxide (PTIO); R,COOH: carboxy-PTIO; R,OCH$_2$CHOOH, carboxymethoxy (CM)-PTIO)

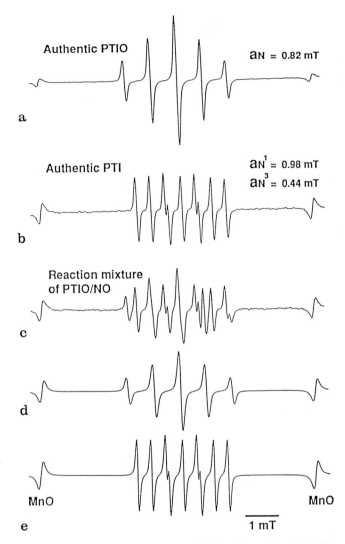

Fig. 3. Electron spin resonance spectra of imidazolineoxyl N-oxide (*PTIO*) reaction product (*PTI*) and reaction mixture of PTIO and nitric oxide (*NO*). **a**, PTIO; **b**, PTI; **c**, spectrum of the reaction mixture of PTIO and NO; **d** and **e**, computer simulation of **c**. (From AKAIKE et al. 1993, with permission)

condition. The results of mass spectroscopy and NMR studies of PTIO and PTI were consistent with the reaction scheme (Fig. 2). The reaction can be quantified either by ESR signal or by spectroscopy of PTIO (λ_{max} 358 nm, ε_{mol} 6130; λ_{max} 562 nm, ε_{mol} 760) and PTI (λ_{max} 432 nm, ε_{mol} 585) (AKAIKE et al. 1993; and our unpublished data). The minimum sensitivity of ESR spectroscopy for NO detection is ~ 0.1 μM.

3 Generation of Nitric Oxide from Rabbit Aorta by Stimulation with Vascular Relaxing Agents and Inhibition of Vascular Relaxation with Imidazolineoxyl *N*-Oxide: Ex Vivo Study

We investigated whether PTIOs react with NO to inhibit its biological function as an EDRF in aortic smooth muscle from rabbits. Stimulation by vascular relaxing agents such as acetylcholine or ATP is known to mediate generation of EDRF, i.e., NO, during this vasorelaxation. Female New Zealand white rabbits weighing 2.5-3 kg, were anesthetized with sodium pentobarbiturate and exsanguinated. The thoracic aorta was removed from the chest and, after removal of excess fat and connective tissue, cut into 5 mm wide rings. The rings were mounted vertically in a 20 ml organ bath filled with Krebs solution, and isometric tension development was recorded with an ink-writing recorder. The medium was maintained at 37 °C and aerated with a mixture of 95% O_2 and 5% CO_2. Tissues were precontracted with 0.15 mM phenylephrine, and acetylcholine- or ATP-induced relaxation was measured, as were the effects of a series of PTIOs, PTI, N^ω-nitro-L-arginine (L-NNA), and N^ω-monomethyl-L-arginine (L-NMMA) on the induced relaxation. The vascular tone of aortic rings was studied in the presence of 3.0 mM indomethacin after phenylephrine-induced contraction of the smooth muscle.

Evidence of NO generation from the vascular rings was obtained using ESR spectroscopy to quantify PTI formed from the reaction with PTIO during stimulation with acetylcholine in the incubation medium. Generation of NO was inhibited in the presence of the NOS inhibitor, L-NMMA (Fig. 4) (see AKAIKE et al. 1993 for details). All three PTIOs used inhibited this NO-related vasorelaxation induced by acetylcholine in a dose-dependent manner (Fig. 5). Similarly, vasorelaxation induced with ATP was also markedly inhibited by all PTIOs (data not shown). In contrast, virtually no inhibitory action was observed against

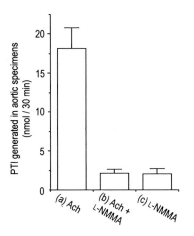

Fig. 4. Conversion of imidazolineoxyl *N*-oxide (*PTIO*) to reaction product (*PTI*) during vasorelaxation induced by acetylcholine (*Ach*). Aortic specimens, 5cm long, from rabbits were incubated with 0.5nM PTIO and Cu, Zn superoxide dismutase (SOD) (500 units/ml) in the presence or absence of 10 nM Ach or 1 mM (*L-NMMA*) in Krebs solution (1.5 ml) for 30 min at 37°C. Error bars indicate mean ±SEM. (From AKAIKE et al. 1993, with permission)

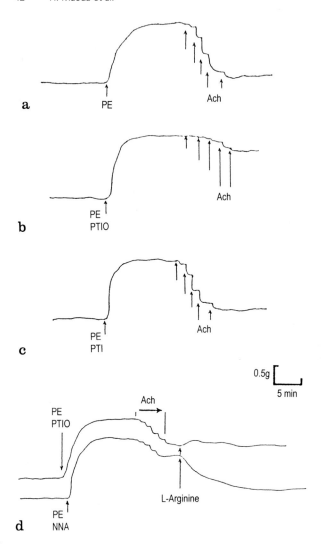

Fig. 5a–d. Effects of imidazolineoxyl *N*-oxide (*PTIO*) on acetylcholine (*Ach*) induced vasorelaxation of rabbit thoracic aorta. Tissues were precontracted with 0.15 n*M* phenylephrine after which Ach was added serially at concentrations of 0.001, 0.03, 0.1, 0.3 and 1.0 m*M* with or without PTIO and *PTI*. **a** Control without PTIO; **b** with 300 μ*M* PTIO; **c** with 300 μ*M* PTI; **d** effect of L-arginine (5 m*M*) on inhibition by 100 μ*M* PTIO (*upper tracing*) or 100 μ*M* *N*$^\omega$-nitro-L-arginine (*NNA*) (*lower tracing*) of Ach-induced vasorelaxation. (From AKAIKE et al. 1993, with permission)

vasorelaxation with PTI, the reaction product of PTIO and NO. The inhibition of the smooth muscle relaxation by PTIOs was not abolished by the addition of L-arginine, the precursor of NO biosynthesis, whereas L-arginine significantly reversed the inhibition of vasorelaxation by L-NNA (Fig. 6). These results indicate that the inhibition of vascular relaxation by PTIOs was not due to direct inhibition of NOS. More specifically, L-arginine analogues such as L-NNA and L-NMMA

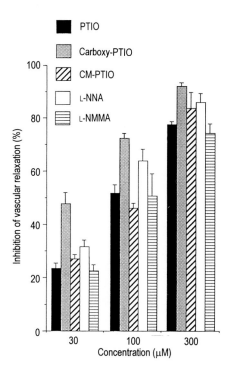

Fig. 6. Effects of concentration of various imidazoline-N-oxide (*PTIO*) derivatives on vascular relaxation of rabbit thoracic aorta. Bioassay conditions were the same as in Fig.5. Data are shown as mean ± SE (*n* = 3-4). (From AKAIKE et al. 1993, with permission)

are competitive inhibitors of NOS and thus the presence of L-arginine can reverse these inhibitory effects (KILBOURN and GRIFFITH 1992). In contrast, PTIO reacts directly with EDRF (NO) generated by cNOS in the endothelium. EDRF-dependent relaxation could be reinduced when acetylcholine was again administered to aortic tissues pretreated with imidazoline compounds for up to 4 hours after stimulation by acetylcholine or ATP, followed by washing out of these reagents. This observation suggests that the inhibitory effect of imidazolineoxyl N-oxide radicals on vasorelaxation did not result from cytotoxicity against endothelial cells of the aorta. The inhibitory potential of PTIOs is comparable to that of L-NMMA and L-NNA (Fig. 6).

4 Therapeutic Effect of Carboxy PTIO on Lipopolysaccharide-Induced Hypotension and Endotoxic Shock, In Vivo

We tested the effect of a water-soluble derivative of PTIO, i.e. carboxy-PTIO, on endotoxic shock rats (Wistar male rats weighing 220–250 g) (YOSHIDA et al. 1994). The lower abdominal aorta was cannulated after abdominal incision to measure mean arterial blood pressure (MABP) and heart rate of the rats, and blood samples were obtained intermittently for ESR analyses to measure the level of carboxy-PTIO and carboxy-PTI. Rats were given bolus intravenous injections of lipopolysaccharide (LPS), *E. coli* 026:B6 (Difco Laboratories, Detroit, MI). Blood pressure and heart rate were measured by a pressure transducer with a recorder unit. Soon after injection of LPS, at a dose of 10 mg/kg, MABP began to decline (Figs. 7, 8). After 3 h, it was lower than 60 mm Hg, and rats started to die. In contrast, when carboxy-PTIO was slowly infused via the left jugular vein, beginning at 90 min at 6 ml/h (1.7 mg/min per kg) and continuing for 60 min the decrease in MABP was inhibited at a carboxy-PTIO dose of more than

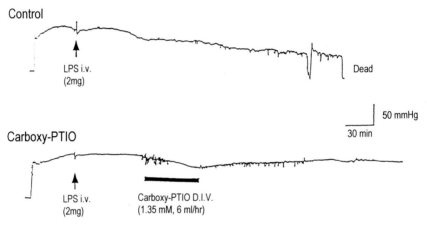

Fig. 7. Effects of carboxy-immidazolineoxyl *N*-oxide (*PTIO*) on lipopolysaccharide (*LPS*)-induced hypotension in rats

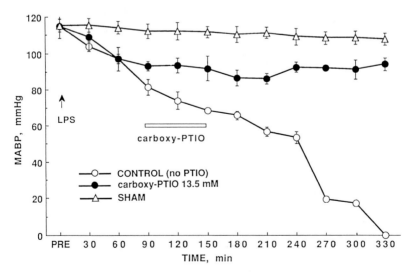

Fig. 8. Therapeutic effect of carboxy-imidazolineoxyl *N*-oxide (*PTIO*) in rats with lipopolysaccharide induced endotoxic shock. MABP, mean arterial blood pressure

0.056 mg/kg per min. At a dose of 0.17 mg/kg per min or higher, carboxy-PTIO infusion completely rescued the endotoxic shock rats.

ESR spectroscopy indicated decreased blood levels of carboxy-PTIO and increased levels of carboxy-PTI in the endotoxic shock rats (not shown). When the infusion was stopped, plasma levels of carboxy-PTIO declined gradually, and both carboxy-PTIO and -PTI were rapidly excreted in the urine (not shown). Total urinary recovery of carboxy-PTI and -PTIO was about 81% in 3 h (not shown).

These results indicate that NO is responsible for the hypotension induced by endotoxin and that carboxy-PTIO has potential therapeutic value against endotoxic shock in animals.

5 Vascular Permeability of Solid Tumors Mediated by Nitric Oxide and Inhibition by Imidazolineoxyl N-Oxide

Tumor tissues have unique vascular characteristics, including extensive angiogenesis (FOLKMAN 1990), which usually results in hypervasculture, irregular morphology and defective architecture, and enhanced permeability (MATSUMURA and MAEDA 1986; SUZUKI et al. 1987; MAEDA and MATSUMURA 1989; SKINNER et al. 1990; MAEDA 1991; MAEDA et al. 1992). This enchanced permeability is relevant in the selective delivery of macromolecules and lipids to tumor tissues, since such tissue exhibits much slower return of these substances into the lymphatics (IWAI et al. 1984; MATSUMURA and MAEDA 1986; MAEDA 1991; MAEDA et al. 1992). We have termed this phenomenon the "enhanced permeability and retention" (EPR) effect of solid tumors (MATSUMURA and MAEDA 1986; MAEDA and MATSUMURA 1989; MAEDA 1991; MAEDA et al. 1992).

The enhanced vascular permeability of solid tumors, which facilitates tumor growth and perhaps metastasis, is now known to be mediated by a number of factors, including tumor vascular permeability factor (SENGER et al. 1983), bradykinin and hydroxyprolyl[3] bradykinin (MAEDA et al. 1988; MATSUMURA et al. 1988, 1991), tumor necrosis factor and interleukin-2 (ETTINGHAM et al. 1988), and others. However, there appears to be no report on the effect of NO on vascular permeability of tumor tissues. Here we present evidence that vascular permeability is indeed mediated by NO (MAEDA et al. 1994).

Figure 9 shows the levels of Evans blue dye relased from sarcona 180 (S-180) solid tumors in ddY mice in the presence or absence of PTIO. PTIO in an oil formulation (0.5 ml, 43 mM) was administered orally and plasma concentrations were determined by ESR spectroscopy. Since plasma levels can be maintained for about 2.5 h, PTIO was administered every 2 h, and Evans blue (0.2% in physiological saline, 0.2 ml) was injected intravenously into tumor-bearing mice. Tumors were more than 3 mm but less than 7 mm in diameter. The results clearly show significant suppression of extravasation by PTIO in two different size groups (Fig. 9).

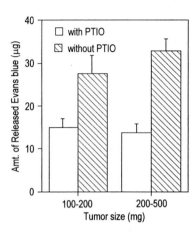

Fig. 9. Inhibitory effect of imidazolineoxyl *N*-oxide (*PTIO*) on vascular permeability in solid tumors. *Bar* indicate SD (*n* = 6-8)

We also tested the extravasating effect of NO in guinea pig skin, in which 0.05 ml of an oil formulation of NO diluted to different concentrations was injected intradermally in animals that had already received Evans blue intravenously. Dye extraction was performed with formamide at 60 °C for 48 h, followed by spectroscopic quantification (620 n*M*). The results showed significant extravasation by NO, although the potency of NO is 1/100–1/1000 of bradykinin (not shown). Extravasation caused by bradykinin was effectively suppressed by PTIO given intraperitoneally (MAEDA et al. 1994).

Data suggest that bradykinin stimulates the release of EDRF (NO) (PALMER et al. 1987; KELM et al. 1988; MYERS et al. 1992; TADJKARIMI et al. 1992). Our data show that NO plays a crucial role in extravasation in tumor tissue, and bradykinin may be an initiating trigger of this event, as discussed previously (MATSUMURA et al. 1988, 1991; MAEDA et al. 1988; MAEDA and MATSUMURA 1989). Further experiments will clarify whether the level of iNOS and for cNOS is elevated in tumor tissue.

6 Microbiocidal Activity of Nitric Oxide Using Imidazolineoxyl *N*-Oxide

It has been suggested that NO plays an important role in the killing of invading bacteria by macrophages and polymorphonuclear cells (PMR) since their NO generation parallels that of oxygen radicals and cell killing (HIBBS et al. 1987; DRAPIER 1991; STUEHR and NATHAN 1991). However, it remains controversial whether NO itself can kill bacteria. We therefore, investigated the effect of NO on *Cryptococcus neoformans* and *Staphylococcus aureus* 209P using PTIO (YOSHIDA et al. 1993).

Staphylococcus aureus was routinely cultured in tryptosoy broth overnight and washed twice with phosphate-buffered saline (PBS; pH 7.3). Cultures were

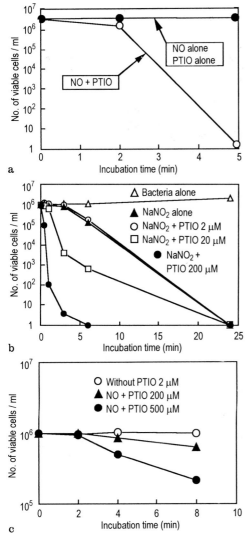

Fig. 10a–c. Effect of imidazolineoxyl N-oxide (*PTIO*) on nitric oxide (*NO*)-induced micro-biocidal action against **a** *Staphylococcus aureus* and *Cryptococcus neoformans*; **b,c** Bactericidal action of NO and PTIO. *S. aureus* (3×10^6 cells/ml) was reacted with NO (150 nM) in 0.01 M PBS (pH 7.3) in the presence or absence of 857 mM PTIO at room temperature. **b** Bactericidal effect of NO generated in a system forming nitrous acid. *C. neoformans* (serotype D) was added to a reaction mixture containing 1 mM $NaNO_2$ in 80 mM succinate buffer (pH 4.0) in the presence or absence of PTIO, followed by incubation for 24 h at room temperature. Number of viable fungal cells was quantitated by colony-forming assay on Sabouraud agar plates. Data represents the mean of triplicate plates. **c** Effects of authentic NO on *C. neoformans* serotype D in solutions at pH 4.0. NO (gas)-saturated solution was added to the suspended *C. neoformans* cells in 200 mM succinate buffer (pH 4.0) and the number of viable cells was measured as in **a**. No cell killing was seen at pH 7.0 and the A serotype showed a trend similar to that of the serotype D (YOSHIDA et al. 1993, with permission)

suspended in PBS and incubated with an aliquot of NO-saturated solution in the presence or absence of PTIO for different time periods (Fig. 10a). Cell suspensions (1×10^6 cells/ml) of *C. neoformans* were prepared in 0.15 M NaCl with varying concentrations of $NaNO_2$ and the suspensions were brought to pH 4.0 by the addition of succinic acid (Fig. 10b). At this pH, NO is known to be released from nitrite. *C. neoformans* was exposed to an NO solution (Fig. 10c) and the effect of PTIO was examined in the same manner as for *S. aureus*. In these assays PTIO alone did not show any cytotoxic activity. The number of viable cells in each reaction system was determined by colony-forming assay.

These results showed that NO alone did not exhibit significant microbiocidal

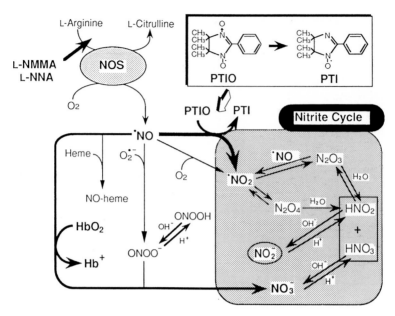

Fig. 11. Interaction of nitric oxide (*NO*) and imidazolineoxyl *N*-oxide (*PTIO*), and various NO intermediates in the nitrite cycle. (From YOSHIDA et al. 1993)

activity against either *S. aureus* or *C. neoformans*. Instead PTIO together with NO showed substantial cytotoxicity. Based on the known chemistry of NO and PTIO (Fig. 11), we speculate that one or more NO metabolites, such as NO_2 (N_2O_4) and N_2O_3, may be toxic but not NO per se. Another derivative, peroxynitrite [OONO], which is formed via the reaction of NO with O_2^- might be also a candidate responsible for NO-dependent microbiocidal action in vivo.

7 Concluding Remarks

The PTIO derivatives are a new class of NO scavenger. The chemistry of radical-radical reactions was fully demonstrated and the product analyses, including mass spectroscopy and NMR studies (not shown here), confirmed this notion (AKAIKE et al. 1993).

   All arginine derivatives may have some drawbacks, especially in vivo. Heme derivatives are also known to react rapidly with NO, but they may also liberate NO and thus may not necessarily be a class of scavenger (RIBIEIRO et al. 1993). In this respect thiol-NO adducts may have the same problems as heme-NO adducts (STAMLER et al. 1992)

   Identity between EDRF and NO was fully established in our study in vitro, ex vivo and in vivo systems using PTIO, and these results are consistent with those obtained with the L-arginine analogue L-NMMA, i.e., PTIO can prevent endotoxin-induced hypotension and shock and can rescue rats from lethal dose of endotoxin (YOSHIDA et al. 1994).

As was mentioned briefly, we found no real evidence that NO is microbiocidal; rather the reaction product(s) of NO and PTIO, e.g. NO_2 or N_2O_4, seems to exhibit more microbiocidal activity (YOSHIDA et al. 1993)

Using a solid tumor model in mice, we demonstrated that the vascular permeability enhancing activity of solid tumors appears to be mediated by NO, because PTIO formulated for slow release significantly suppressed the vascular permeability of Evans blue dye. In conjuction with this finding, PTIO also suppressed the vascular permeability enhancement induced by bradykinin in normal guinea pig skin (MAEDA et al. 1994). Our results indicate that the action of bradykinin on blood vessels is, at least in part, mediated by NO. PTIO will be a very useful reagent for the analysis of NO-mediated activities both in vitro and in vivo.

References

Akaike T, Yoshida M, Miyamoto Y, Sato K, Kohno M, Sasamoto K, Miyazaki K, Ueda S, Maeda H (1993) Antagonistic action of imidazolineoxyl N-oxides against endothelium-derived relaxing factor/. NO through a radical reaction. Biochemistry 32: 827–832

Bredt DS, Snyder SH (1990) Isolation of nitric oxide synthetase, a calmodulin-requiring enzyme. Proc Natl Acad Sci USA 87: 682–685

Drapier JC (1991) L-Arginine-derived nitric oxide and the cell mediated immune response. Res Immunol 142: 551–602

Ettinghausen SE, Puri RJ, Rosenberg SA (1988) Increased vascular permeability in organs mediated by the systematic administration of lymphokine-activated killer cells and recombinant interleukin-2 in mice. J Natl Cancer Inst, 80: 177–187

Folkman J (1990) What is the evidence that tumors are angiogenesis dependent? J Natl Cancer Inst 82:4–6

Gatley SJ, Sherratt HSA (1976) The effects of diphenyleneiodonium on mitochondrial reactions: Relation of binding of diphenylene [[125]]iodonium to mitochondria to the extent of inhibition of oxygen uptake. Biochem J 158: 307–315

Gross SS, Levi R (1992) Tetrahydrobiopterin synthesis: an absolute requirement for cytokine-induced nitric oxide generation by vascular smooth muscle. J Biol Chem 267: 25722–25729

Hibbs JB Jr, Taintor RR, Vartin Z (1987) Macrophage cytotoxicity: role for L-arginine deiminase and iminonitrogen oxidation to nitrite. Science 235: 473–476

Iwai K, Maeda H, Konno T (1984) Use of oily contrast medium for selective drug targeting to tumor: enhanced therapeutic effect and X-ray image. Cancer Res 44: 2114–2121

Kanner J, Harel S, Granit R (1992) Nitric oxide, an inhibitor of lipid oxidation by lipoxygenase, cyclooxygenase and hemoglobin. Lipids 27: 46–49

Kelm M, Feelisch M, Spahr R, Pipper M, Noack E, Schrader J (1988) Quantitative and kinetic characterization of nitric oxide and EDRF released from cultured endothelial cells. Biochem Biophys Res Commun 154: 236–244

Kilbourn RG, Griffith OW (1992) Overproduction of nitric oxide in cytokine-mediated and septic shock. J Natl Cancer Inst 84: 827–831

Maeda H, Matsumura Y (1989) Tumoritropic and lymphotropic principles of macromolecular drugs. Crit Rev Ther Drug Carrier Syst 6: 193–210

Maeda H (1991) SMANCS and polymer-conjugated macromolecular drugs: advantages in cancer chemotherapy. Adv Drug Deliv Rev 6: 181–202

Maeda H, Matsumura Y, Kato H (1988) Purification and identification of [hydroxyprolyl[3]]-bradykinin in ascitic fluid from a patient with gastric cancer. J Biol Chem 263: 16051–16056

Maeda H, Seymour LW, Miyamoto Y (1992) Conjugates of anticancer agents and polymers: Advantages of macromolecular therapeutics in vivo. Bioconj Chem 3: 351–362

Maeda H, Noguchi Y, Sato K, Akaike T (1994) Enhanced vascular permeability in solid tumor is mediated by nitric oxide and inhibited by both new nitric oxide scavenger and nitric oxide synthase inhibitor. Jpn J Cancer Res 85: 331–334

Matsumura Y, Maeda H (1986) A new concept of macromolecular therapeutics in cancer chemotherapy: mechanism of tumoritropic accumulation of proteins and anti tumor agent SMANCS. Cancer Res 46: 6387–6392

Matsumura Y, Kimura M, Yamamoto T, Maeda H (1988) Involvement of kinin-generating cascade in enhanced vascular permeability in tumor tissue. Jpn J Cancer Res 79: 1327–1334

Matsumura Y, Maruo K, Kimura M, Yamamoto T, Konno T, Maeda H (1991) Kinin-generating cascade in advanced cancer patients and in vitro study. Jpn J Cancer Res 82: 732–741

Moncada S, Palmer RMJ, Higgs EA (1992) Nitric oxide: physiology, pathophysiology, and pharmacology. Pharmacol Rev 43: 109–142

Myers PR, Guerra R Jr, Harrison DG (1992) Release of multiple endothelium-derived relaxing factors from porcine coronary arteries. J Cardiovasc Pharmacol. 20: 392–400

Palmer RMJ, Ferrige AG, Moncada S (1987) Nitric oxide release accounts for the biological activity of endothelium-derived relaxing factor. Nature 327: 524–526

Ribieiro JMC, Hazzard JMH, Nussenzveig RH, Champagene DE, Walker FA (1993) Reversible binding of nitric oxide by a salivary heme protein from a bloodsucking insect. Science 260: 539–541

Sakai N, Kaufman S, Milstien S (1993) Tetrahydrobiopterin is required for cytokine-induced nitric oxide production in a murine macrophage cell line (RAW 264). Mol Pharmacol 43: 6–10

Senger DR, Galli SJ, Dvorak AM, Perruzzi CA, Harvey VS, Dvorak HF (1983) Tumor cells secrete a vascular permeability factor that promotes accumulation of ascitic fluid. Science 219: 983–985

Skinner SA, Tutton PJM, O'Brein E (1990) Microvasculature architecture of experimental colon tumor in rats. Cancer Res 50: 2411–2417

Stamler JS, Jaraki O, Osborne J, Simon DI, Keaney J, Vita J, Singel D, Valeri CR, Loscalzo J (1992) Nitric oxide circulates in mammalian plasma primarily as an S-nitroso adduct of serum albumin. Proc Natl Acad Sci USA 89: 7674–7677

Stuehr DJ, Nathan CF (1991) Nitric oxide: a macrophage product responsible for cytostasis and respiratory inhibition in tumor target cells. J Exp Med 169: 1543–1555

Stuehr DJ, Fasehun OA, Kwon NS, Gross SS, Gonzales JA, Levi R, Nathan CF (1991) Inhibition of macrophage and endothelial cell nitric oxide synthase by diphenyleneiodonium and its analogs. FASEB J 5: 98–103

Suzuki M, Takahashi T, Sato T (1987) Medial regression and its functional significance in tumor supplying host arteries. Cancer 59: 444–450

Tadjkarimi S, O'Neil GS, Luu TN, Allen SP, Schyns CJ, Chester AH, Yacoub MH (1992) Comparison of cyclic GMP in human internal mammary artery and saphenous vein: implications for coronary artery bypass graft patency; Cardiovasc Res 26: 297–300

Yoshida K, Akaike T, Doi T, Sato K, Ijiri S, Suga M, Ando M, Maeda H (1993) Pronounces enhancement of NO-dependent antimicrobial action by an NO-oxidizing agent, imidazolineoxyl N-oxide. Infect Immun 61: 3552–3555

Yoshida M, Akaike T, Wada Y, Sato K, Ikeda K, Ueda S, Maeda H (1994) Therapeutic effects of imidazolineoxyl N-oxide against endotoxin shock through its direct nitric oxide-scavenging activity. Biochem Biophys Res Commun 202: 930–932

The Role of Nitric Oxide in the Pathogenesis of Virus-Induced Encephalopathies

B. Dietzschold

1 Introduction

The mechanism by which virus infections of the central nervous system (CNS) cause neuronal damage are understood in only a few viral CNS diseases in which there is evidence that the virus directly destroys its target cells. In many other cases of virus-induced encephalopathy, the virus does not directly destroy neural tissue but causes indirect damage by altering neuronal functions (Fu et al. 1993; Lipkin et al. 1988a,b) or by triggering cell-mediated responses within the CNS (Byrne and Oldstone 1984; Carbone et al. 1988; Doherty et al. 1976). Soluble factors such as proinflammatory cytokines, proteases, free radicals, and neurotoxins produced by immune cells are thought to play an important role in the process of neuronal destruction (Shankar et al. 1992; Koprowski et al. 1993). It is believed that proinflammatory cytokines in particular are exceptionally important in the process of neuronal destruction (Selmaj 1992). It has been shown that cytokines can have direct cytotoxic effects. For example, the intracisternal challenge of interleukin-1 (IL-1) or tumor necrosis factor (TNF) in rats induced meningitis and blood brain barrier damage (Quagliarello et al. 1991). Recently, attention has focused on the possibility that reactive nitrogen intermediates

Department of Microbiology & Immunology, Center for Neurovirology, J.A.H. # 459, Thomas Jefferson University, 1020 Locust Street, Philadelphia, PA 19107, USA generated by nitric oxide synthases (NOS) directly damage neural tissue. Both the constitutive form of NOS (cNOS), and the inducible calcium-independent NOS (iNOS) are thought to play a role in the process of neuronal damage (Koprowski et al. 1993; Murphy et al. 1993). NO generated by cNOS in neurons has been proposed to initiate excitotoxicity through activation of guanylate cyclase (Dawson et al. 1991). However, the involvement of cyclic GMP in NO-mediated neurotoxicity is questionable since inhibitors of guanylate cyclase failed to protect neuron cultures from the toxic effect of NMDA or the NO donor sodium nitroprusside (Lustig et al. 1992). In contrast, increasing evidence suggests a role for activated microglia or infiltrating macrophages in neuronal damage and white matter pathologies by a TNF-and iNOS-dependent process. It has been shown in several neurological diseases that the expression of proinflammatory cytokines, such as IL-1, TNF, and interferon (IFN)-γ, which can prime macrophages and microglial cells for the production of reactive nitrogen intermediates (Ding et al. 1988; Nathan 1992), correlates with the degree of inflammatory lesions in the CNS and with the severity of neurological signs (Koprowski et al. 1993; Muprhy et al. 1993). Furthermore, the observation that iNOS mRNA is present in the brains of virus-infected rats or rats with experimental allergic encephalitis (EAE) (Koprowski et al. 1993), is very suggestive for the involvement of NO in neuronal damage. To further investigate the potential pathophysiological role of NO in virus-induced neuropathogenesis, we have determined the kinetics of iNOS and cNOS expression in rabies virus and Borna disease virus (BDV)-infected rat brains. Rabies virus and BDV were chosen in this study because both viruses are highly neurotropic. Infection with BDV results in a persistent CNS infection which is characterized by massive infiltration of inflammatory cells. Our results show that iNOS expression which is not detectable in normal brain tissue is strongly up-regulated in the brain of rats infected with BDV or rabies virus and levels of iNOS activity correlate with clinical severity. In contrast, expression of neuronal cNOS, which is detectable in normal brain, is significantly decreased during BDV and rabies virus infection.

2 Kinetics of Expression of Tumor Necrosis Factor-α and Inducible Nitric Oxide Synthase mRNA in Rat Brains Following Borna Disease Virus Infection

Since it has been proposed that TNF-α, TNF-β and IFN-γ can prime macrophages for production of inducible reactive nitrogen intermediates, we compared the kinetics of BDV-induced changes in the mRNA levels of iNOS and TNF-α. The results showed that iNOS and TNF-α mRNAs, normally nondetectable in the brain, were present in the brain after infection with BDV (Zheng et al. 1993). Although reverse transcriptase-polymerase chain reaction analysis was used, which yields qualitative rather than quantitative results, highest levels of iNOS mRNA were detected 21 days postinfection (p.i.) when animals exhibited neurological signs such as seizures, convulsions, and tremore. Expression of TNF-α mRNA was at a maximum 17 days p.i. and decreased at 21 days p.i. (ZHENG et al. 1993) indicating that the expression of TNF-α precedes that of iNOS. This result is consistent with the notion that certain proinflammatory cytokines such as TNF-α, IL-1 and IFN-γ might effect CNS damage by stimulating inflammatory cells to produce oxygen and nitrogen intermediates.

3 Localization of Inducible Nitric Oxide Synthase Expression in the Borna Disease Virus-Infected Brain

In situ hybridization experiments revealed that iNOS mRNA is expressed in brain areas that are preferentially infected by BDV, such as the hippocampus and basolateral parts of the cortex. However, unlike BDV RNA, which is evenly distributed among neuronal cell layers, iNOS mRNA exhibited a patchy distribution pattern. No iNOS mRNA could be detected in brains of normal rats.

iNOS producing cells in the BDV infected brain were identified by immunohistochemistry using antibodies against iNOS, the macrophage/monocyte marker ED1, and the astroglial marker GFAP. This analysis revealed a generalized astrogliosis accompanied by massive macrophage infiltration in brains of rats with Borna disease (E. Weihe, unpublished). However, cells expressing GFAP and those expressing ED1 were distinctly segregated. The distribution pattern of iNOS-positive cells was very similar to that of ED1-positive cells, and the staining for iNOS was confirmed to the same type of cells as that stained for ED1. Conversely, iNOS-positive cells were absent in areas of intensive GFAP staining. The segregation of areas containing GFAP-stained cells and areas with an accumulation of cells stained for ED1 and iNOS may reflect the damaging effect of NO released by activated macrophages or activated microglia. It is likely that astrocytes, neurons, and oligodendrocytes located in areas where iNOS-positive cells are present are functionally impaired or even destroyed. Furthermore, the observation that cells intensely stained for iNOS are present in areas where neurons were still morphologically intact, whereas iNOS expression is less prominent in areas of massive tissue destruction (ZHENG et al. 1993), suggests that the generation of NO by iNOS is an early event in the process of neurological damage.

4 Effect of Viral Infections on Neuronal Constitutive Nitric Oxide Synthase in the Brain

Excessive amounts of NO in neurons generated by cNOS have been suggested to be involved in neurodegenerative processes such as excitotoxicity (DAWSON et al. 1991). However, determination of NOS activities in the brain indicated that activity of cNOS in the cerebrum decreased after infection with BDV (Table 1) or rabies virus (AKAIKE et al. 1994). In BD, the decrease in cNOS activity which could already be observed 16 days after infection, declined to 30% of the activity found in normal rat brain on day 22 after infection.

Western blot analysis using an antibody specific for cNOS revealed a reduction in neuronal cNOS in the brain after infection with BDV. The observation that cNOS mRNA also decreased in a time-dependent manner after BDV infection indicates that the decrease in neuronal cNOS activity is most likely due to a decrease in the expression of neuronal cNOS mRNA.

The decrease in cNOS expression could be the result of extensive neuronal loss that might occur in virus-infected brains. However, the activity of choline acetyl; transferase remained unchanged after BDV or rabies virus infection. Furthermore, expression of several immediate-early response genes, such as c-fos, jun B, egr-1, and the late response gene encoding proenkephalin is markedly increased in BDV-infected neurons of the cerebral cortex and the hippocampus (FU et al. 1993). Therefore, it is unlikely that the decrease in cNOS activity is caused by a massive neuronal destruction. Selective damage of nitroxergic neurons is also unlikely because recent observations have shown that, in lesions of patients with Alzheimer's disease or in brains with experimentally induced ischemia, nitroxergic neurons are spared from destruction (HYMAN et al. 1992; UEMURA et al. 1990). Since BDV or rabies virus can induce the expression of transcription factors, it is possible that the decrease of cNOS expression is the

Table 1. Inducible and Constitutive Nitric Oxide Synthase in Normal and Borna Disease Virus Infected Rat Brain[a]

Days Postinfection	Enzyme activity (μM/20 min/g tissue)	
	iNOS	cNOS
Controls	0	28.9 ± 1.1
16	0.39 ± 0.1	23.4 ± 1.2
18	1.92 ± 1.0	19.2 ± 0.7
19	1.78 ± 0.7	18.8 ± 1.0
20	2.17 ± 0.4	14.7 ± 1.4
22	1.13 ± 0.5	8.7 ± 1.4

[a] iNOS, inducible nitric oxide synthase; cNOS, constitutive nitric oxide synthase.
NOS activity was measured radiochemically as described by BREDT and SNYDER (1989). To differentiate the activity of iNOS from cNOS, the calmodulin inhibitor trifluroperazine (final concentration: 100 μM) was added to the assay in vitro

result of the down-regulation of cNOS mRNA transcription by some of these transcription factors, although it cannot be excluded that post-translational modifications of NOS such as phosphorylation (BREDT et al. 1992) or inhibition of NOS by soluble factors such as IL-4 and transforming growth factor (TGF)-β (NATHAN 1992) contribute to the decrease of cNOS activity.

NO generated by neuronal cNOS is thought to be an important neuro-transmitter that plays a role in long-term potentiation and long-term depression (BLISS and COLLINGRIDGE 1993). Since BDV and rabies virus replicate preferentially in the hippocampal formation, it is possible that the behavioral abnormalities seen in these virus infections could be partially due to the suppression of neuronal cNOS.

5 Effect of Nitric Oxide Synthase Inhibitors on Borna Disease Virus-Induced Encephalopathy

To determine whether NO, generated by macrophages, plays a major role in the development of neurological symptoms and neuropathology of BDV infection, rats were treated with NOS inhibitors such as L-N-monomethylarginine (L-NMMA) and L-nitroargininemethylester (L-NAME) (Table 2). However, none of these inhibitors exerted a therapeutic effect. Treatment with L-NMMA resulted in an exacerbation of clinical signs. These data suggest that NO by itself does not play a pivotal role in the neuropathogenesis of viral CNS infections. On the contrary, the exacerbation of clinical signs after treatment with NOS inhibitors suggests that reduced NO production, caused by the decrease in neuronal cNOS activity in BDV-infected rats, may actually contribute to the disease process.

Table 2. Effect of Nitric Oxide Synthase Inhibitors and Scavenger on Borna Disease Virus Infected Rats

Treatment[a]	Rats with clinical signs 10 days after treatment	Survivor rate	Loss of body weight 25 days after infection (gm)[b]
Control	4/5	5/5	11.26 ± 4.99
L-NAME	5/5	4/5	18.95 ± 5.54
L-NMMA	5/5	4/5	26.17 ± 2.81

[a] Rats were treated beginning at day 15 postinfection (p.i.) for 10 days. All compounds were delivered via i.p. implanted Alzet pumps; L-NAME and L-NMMA at a rate of 140 µ/h/kg. Control animals received PBS via i.p. implanted Alzet osmotic pumps.
[b] Data are mean ± SE of four or five rats that survived 25 days after Borna disease virus infection. Loss of body weight in control and L-NMMA-treated rats differed significantly ($p < 0.05$), two-tailed t-test.

References

Akaike T, Weihe E, Schaeffer MK-H, Fu ZF, Zheng YM, Vogel WH, Schmidt HHHW, Koprowski H, Dietzschold B (1994) Effect of neurotropic virus infection on neuronal and inducible nitric oxide synthase activity in rat brain. J Neurovirol (in press)

Bliss TVP, Collingridge GL (1993) A synaptic model of memory: long-term potentiation in the hippocampus. Nature 361: 31–39

Bredt DS, Snyder SH (1989) Nitric oxide mediates glutamate-linked enhancement of cGMP levels in the cerebellum. Proc Natl Acad Sci USA 86: 9030–9033

Bredt DS, Ferris CD, Snyder SH (1992) Nitric oxide synthase regulation sites: phosphorylation by cyclic AMP-dependent protein kinase, protein kinase C, and calcium/calmodulin protein kinase; identification of flavin and calmodulin binding sites. J Biol Chem 267: 10976–10981

Byrne JA, Oldstone MBA (1984) Biology of cloned cytotoxic T lymphocytes specific for lymphocytic choriomeningitis: clearance of virus in vivo. J Virol 51: 682–686

Carbone KM, Duchala CS, Narayan O (1988) Borna disease: an immunopathologic response to viral infection in the CNS. Ann N.Y Acad Sci 540: 661–662

Dawson VL, Dawson TM, London ED, Bredt DS, Snyder SH (1991) Nitric oxide mediates glutamate neurotoxicity in primary cortical cultures. Proc Natl Acad Sci USA 88: 6368–6371

Ding AH, Nathan CF, Stuehr DJ (1988) Release of reactive nitrogen and oxygen intermediates from mouse peritoneal macrophages. Comparison of activating cytokines and evidence for independent production. J Immunol 141: 2407–2412

Doherty PC, Dunlop MBC, Parish CR, Zinkernagal RM (1976) Inflammatory process in murine lymphocytic choriomeningitis is maximal in H-2K or H-2D compatible interactions. J Immunol 117: 187–190

Fu ZF, Weihe E, Zheng YM, Schäffer MK-H, Sheng H, Corisdeo S, Rauscher FJIII, Koprowski H, Dietzschold B (1993) Differential effects of rabies and Borna disease viruses on immediate-early- and late-response gene expression in brain tissue. J Virol 67: 6674–6681

Hyman BT, Marzloff K, Wenniger JJ, Dawson TM, Bredt DS, Snyder SH (1992) Relative sparing of nitric oxide synthase-containing neurons in the hippocampal formation in Alzheimer's disease. Ann Neurol 32: 818–820

Koprowski H, Zheng YM, Heber-Katz E, Fraser N, Rorke L, Fu ZF, Hanlon CA, Dietzschold B (1993) In vivo expression of inducible nitric oxide synthase in experimentally induced neurologic diseases. Proc Natl Acad Sci USA 90: 3024–3027

Lipkin WI, Battenberg ELF, Bloom FE, Oldstone MBA (1988a) Viral infection of neurons can depress neurotransmitter mRNA levels without histologic injury. Brain Res 451: 333–339

Lipkin WI, Carbone KM, Wilson MC, Duchala CS, Narayan O, Oldstone MBA (1988b) Neurotransmitter abnormalities in Borna disease. Brain Res 475: 336–370

Lustig HS, von Brauchitsch KL, Chan J, Greenberg DA (1992) Cyclic GMP modulators and excitotoxic injury in cerebral cortical cultures. Brain Res 577: 343–346

Murphy S, Simons ML, Agullo L, Garcia A, Feinstein DL, Galea E, Reis DJ, Minc-Golomb D, Schwartz JP (1993) Synthesis of nitric oxide in CNS glial cells. Trends Neurosci 16: 323–328

Nathan C (1992) Nitric oxide as a secretory product of mammalian cells. FASEB J 6: 3051–3064

Quagliarello VJ, Wispelwey B, Long WJ Jr, Scheid WM (1991) Recombinant human interleukin-1 induces meningitis and blood-brain barrier injury in the rat. J Clin Invest 87: 1360–1366

Selmaj KW (1992) The role of cytokines inflammatory conditions of the central nervous system. Semin Neurosci 4: 221–225

Shankar V, Kao M, Hamir AN, Sheng H, Koprowski H, Dietzschold B (1992) Kinetics of virus spread and changes in levels of several cytokine mRNAs in brain after intranasal infection of rats with Borna disease virus. J Virol 66: 992–998

Uemura Y, Kowall NW, Beal MF (1990) Selective sparing of NADPH-diaphorase-somatostatin-neuropeptide Y neurons in ischemic bergil striatum. Ann Neurol 27: 620–625

Zheng YM, Schäffer MK-H, Weihe E, Sheng H, Corisdeo S, Fu ZF, Koprowski H, Dietzschold B (1993) Severity of neurological signs and degree of inflammatory lesions in the brains of rats with Borna disease correlate with the induction of nitric oxide synthase. J Virol 67: 5786–5791

The Role of Peroxynitrite
in Nitric Oxide-Mediated Toxicity

J.P. CROW and J.S. BECKMAN

1 Introduction

Nitric oxide is a hydrophobic gas with chemical properties that make it uniquely suited as both an intra- and intercellular messenger. It is a relatively stable, charge-neutral radical with a strong tendency to interact and react with other species possessing unpaired electrons such as superoxide, iron, and molecular oxygen. Thus, it has several potential toxic mechanisms; actual toxicity is highly dependent on nitric oxide concentration and the particular microenvironment in which it is produced.

Nitric oxide has been reported to inhibit critical iron-sulfur-containing enzymes involved in mitochondrial respiration (WELSH et al. 1991), to inhibit ribonucleotide reductase (KWON et al. 1991; LEPOIVRE et al. 1991), and to damage DNA directly (NGUYEN et al. 1992; WINK et al. 1991). In addition, nitric oxide has been implicated in reoxygenation injury following ischemia (JAESCHKE et al. 1992; Masini et al. 1991), glutamate-mediated neuronal toxicity (CAZEVIEILLE et al. 1993;

University of Alabama at Birmingham, Department of Anesthesiology, 619 19th Street South, THT 958, Birmingham, Alabama 35233, USA

Dawson et al. 1993), inflammation (Berrazueta et al. 1990; Kroncke et al. 1991; Mulligan et al. 1991), graft host disease (Hoffman et al. 1992; Garside et al. 1992; Langrehr et al. 1992), and as a major arm of host defense against viruses, bacteria, and other intracellular parasites (Green et al. 1990a,b). In many cases these findings were based on the ability of either nitric oxide synthase inhibitors or nitric oxide scavengers like hemoglobin to protect against a particular pathological change or, in the latter case, to inhibit parasite killing by activated macrophages. Conversely, compounds which release nitric oxide typically cause or exacerbate the change. The most reasonable extrapolation from these studies is that nitric oxide is directly toxic.

Many investigators readily accept the notion that nitric oxide is toxic, given its reputation as a reactive radical with a short biological half-life. However, reactive is not synonomous with toxic. Nitric oxide is known to react with molecular oxygen to give nitrogen dioxide (the orange-brown gas in smog), which is a strong oxidant and potent toxin in any biological system, but significant nitrogen dioxide formation occurs only at nitric oxide concentrations far in excess of those seen in vivo. The reaction of nitric oxide with superoxide occurs very rapidly even at physiological concentrations of nitric oxide and offers a more plausible explanation for both the short biological half-life of nitric oxide and its participation in toxic reactions.

An effort will be made in this chapter to examine the chemistry of nitric oxide, with particular attention to reactions which can and do occur in biological systems and those which are unlikely to occur. Major emphasis will be given to the reaction of nitric oxide with superoxide in vivo to form peroxynitrite, a potent oxidant with multiple reaction pathways. A strong case will be made for the participation of peroxynitrite in nitric oxide-mediated toxicity. Existing theories of oxygen radical-mediated toxicity will first be examined and attempts will be made to account for older findings in light of newer hypotheses.

2 The Superoxide Theory of Toxicity

Molecular oxygen is readily reduced by one electron to give superoxide anion. The term "superoxide" was originally coined to describe its unusual electronic configuration rather than its reactivity (Pauling 1979). However, for many years, superoxide was taken to mean "superoxidizing". Reputable biochemistry textbooks still refer to superoxide anion as a "highly reactive and destructive radical". At physiological pH, superoxide has little potential to oxidize other molecules since this would require placing a second negative charge on superoxide (Sawyer and Valentine 1981; Baum 1984). Neutralization of the first negative charge by protonation facilitates superoxide acting as an oxidant; however, the pK_a for superoxide is 4.8 (Bielski 1978). Thus, superoxide acts as an oxidant only in a more acidic environment or when the molecule under attack is a good one-electron donor such as reduced iron in Fe (II)-sulfur centers. Iron (III) is readily reduced by superoxide; superoxides' ability to carry out the one-electron reduction of iron (III) in cytochrome c serves as the basis for a frequently used assay (McCord and Fridovich 1969).

Superoxide dismutase (SOD) affords protection against oxidant stress and ischemic injury in many model systems (Flaherty and Weisfeldt 1988; Zweier et al. 1987). The mechanism by which SOD protects remains unclear since few specific biological targets for superoxide have been identified. Superoxide appears to reversibly inactivate mitochondrial Fe (II)-sulfur enzymes in bacteria (Gardner and Fridovich 1991a,b); the percent of dynamically inactivated aconitase has seen used to estimate the steady-state concentration of superoxide (Gardner and Fridovich 1992). However, the contribution of this ongoing inactivation/ reactivation to overt cellular toxicity is not clear. To account for the effects of SOD in the absence of clear direct toxicity of superoxide, the superoxide-driven Fenten reaction was proposed whereby superoxide reduces Fe^{+3} to Fe^{+2}, which, in turn, reduces hydrogen peroxide resulting in the formation of the strongly oxidizing hydroxyl radical (see below). Numerous problems exist with this proposal, not the least of which is its inability to account for nitric oxide-dependent toxicity.

3 Hydroxyl Radical Is Too Reactive To Be Highly Toxic

For many years, the accepted explanation for superoxide-derived toxicity has been hydroxyl radical production via the iron-catalyzed Haber-Weiss reaction (Haber and Weiss 1934):

$$O_2^- + Fe^{3+} \rightarrow O_2 + Fe^{2+}$$

$$Fe^{2+} + H_2O_2 \rightarrow \cdot OH + {}^-OH + Fe^{3+}$$

In this scheme, superoxide serves primarily as the reductant for iron. This seems unlikely when considering that the steady-state superoxide concentration is in the range of 10–100pM (Gardner and Fridovich 1992) while other reductants like ascorbate are present in much higher concentrations in vivo and also reduce Fe^{3+} to Fe^{2+} quite effectively. On the other hand, if superoxide served primarily as the source of hydrogen peroxide, then SOD would not be protective (since hydrogen peroxide is formed from the reaction of superoxide with SOD). Superoxide may also serve to increase the pool of free iron by releasing it from ferritin (Biemond et al. 1984; Thomas et al. 1985). However, even with a relatively large pool of Fe^{2+}, the rate at which it reduces hydrogen peroxide to produce hydroxyl radical is quite slow (10^3–10^4 $M^{-1}s^{-1}$) (Rush and Koppenol 1989). Thus, the iron-catalyzed Haber-Weiss reaction requires the interaction of three species present in low concentrations in vivo due to efficient scavenging systems (SOD, catalase, ferritin), making the rate of hydroxyl radical formation dependent on the product of three minuscule concentrations which react at relatively low rates.

Even if hydroxyl radical were efficiently formed in vivo, it is simply too reactive to do any specific damage. The ability of hydroxyl radical to react at near diffusion-limited rates with virtually all biological molecules limits its effective diffusion radius to only a few angstroms. Thus, hydroxyl radical would have to be formed directly at critical sites (e.g., active sites of enzymes, bases of DNA, within membranes) to affect significant injury. Many such critical targets are capable of binding iron in vitro thereby promoting hydroxyl radical formation in close proximity. However, the rates of hydrogen peroxide reduction by bound iron can be even lower than with free iron due to nonoptimal coordination geometries. Overall, the slow rates of reactions leading to hydroxyl radical production and the low concentrations of the reactants suggest that other reactions may be important for understanding superoxide toxicity.

4 The New Radical Era—Nitric Oxide as Cellular Messenger

The discoveries of nitric oxide synthases and nitric oxide production in biological systems ushered in a new era of oxygen radical enthusiasm reminiscent of that seen previously with superoxide. A major biological difference between superoxide and nitric oxide is that cells have adapted to tolerate the inescapable production of superoxide, while a variety of isoenzymes have evolved to deliberately and specially produce nitric oxide. At present, there is no evidence for "a nitric oxide dismutase" to eliminate this newly discovered reactive radical. Important questions are naturally raised: what determines the half-life of nitric oxide in biological systems? How are its physiological effects terminated and how is toxicity avoided or minimized? Earlier data suggested that superoxide might be involved.

5 Superoxide Dismutase Increases the Biological Half-Life of Nitric Oxide

Prior to conclusive demonstration that endothelium-derived relaxing factor (EDRF) was indeed nitric oxide, SOD had been found to potentiate or prolong the vasorelaxant effects of EDRF (IGNARRO et al. 1989; HUTCHINSON et al. 1987). This led to the concept of superoxide as an endothelial-dependent contracting factor (EDCF) and later to the notion that superoxide simply detoxifies nitric oxide by a direct radical-radical annihiliation reaction. The concept of detoxification will be discussed later. Here, it is noteworthy that the reaction of superoxide with nitric oxide (HUIE and PADMAJA 1993) is three fold faster than the reaction of superoxide with SOD (KLUG et al. 1972) and 30-fold faster than the reaction of superoxide with the Escherichia coli iron-sulfur enzyme 6-phosphogluconate dehydratase

(GARDNER and FRIDOVICH 1991b). Thus, to date, nitric oxide is the only identified target for superoxide which reacts fast enough and can exist at sufficient concentrations to effectively compete with SOD. The ability of SOD to affect the biological half-life of nitric oxide provides strong evidence that the reaction of nitric oxide with superoxide is significant in vivo.

The reaction between nitric oxide and superoxide is directly analogous to a binary chemical bomb, in which two fairly innocuous substances are mixed to produce a potent toxin. Macrophages and neutrophils, cell types which simultaneously produce nitric oxide and superoxide, appear to utilize this strategy to kill invading parasites; accompanying inflammation of host tissues may also be a direct result of this programmed oxidant attack. Peroxynitrite is a strong oxidant and nitrating agent capable of producing tissue injury by targeting key cellular components. Formation of peroxynitrite in cell systems is inhibited by SOD in a dose-dependent manner and by inhibitors of nitric oxide synthase (ISCHIROPOULOS et al. 1992a). Tissues containing constitutive nitric oxide synthase would be primed during ischemic periods to produce both nitric oxide and superoxide during reoxygenation. Hyperoxic conditions would also promote sumultaneous production. Thus, many of the findings related to tissue injury (and protection) under conditions of oxidant stress, ischemia/reperfusion, and inflammation can be explained on the basis of peroxynitrite formation, without the need to invoke hydroxyl radical formation via sequential, concerted reactions that occur at relatively low rates.

6 The Biological Chemistry of Nitric Oxide

Nitric oxide is a radical by virtue of having one unpaired outer shell electron. However, this is not an inherently unstable state since nitrogen normally possesses an odd number of valence electrons (which is why the nitric oxide radical has no formal charge). Nitric oxide can undergo a one-electron reduction to give nitroxyl anion (NO^-) which can also produce peroxynitrite via a direct reaction with molecular oxygen. However, it seems more likely that nitric oxide would react directly with other radicals or one-electron donors/acceptors such as the iron in heme, particularly considering the abundance of heme-containing proteins. Nitric oxide is unusual in that it reacts with both ferrous and ferric forms of heme iron. Signal transduction is accomplished by the binding of nitric oxide to the heme prosthetic group of guanylate cyclase (reaction 1) which activates the enzyme to produce cGMP

$$GC\text{-}heme\text{-}Fe^{+2} + \cdot NO \rightarrow GC\text{-}heme\text{-}Fe^{+2}\text{-} NO. \tag{reaction 1}$$

The major sink for nitric oxide which diffuses to the vasculature is oxyhemoglobin. This results in the formation of nitrate and methemoglobin (reaction 2)

$$Hb\text{-}Fe^{+2}\text{-} O_2 + \cdot NO \rightarrow Hb\text{-}Fe^{+2}\text{---}OONO \rightarrow Hb\text{-}Fe^{+3} + NO_3^-. \tag{reaction 2}$$

The most abundant, hydrophobic, small molecule containing an unpaired electron with which nitric oxide can readily react is oxygen. Nitric oxide reacts in a 2:1 molar ratio with molecular oxygen to form nitrogen dioxide (reaction 3), the orange gas seen when concentrated nitric oxide is released from a cylinder

$$2 \cdot NO + O_2 \rightarrow 2 \cdot NO_2 .$$ (reaction 3)

The rate of nitrogen dioxide formation via this reaction is 2×10^6 $M^{-2}s^{-1}$ at 25 °C (FORD et al. 1993) and since it requires the interaction of three molecules, the rate is dependent on the square of nitric oxide concentration (reaction 4). This has important biological implications

$$d[NO]/dt = k_3 [NO]^2 [O_2] \quad \text{where } k_3 = 4 \times 10^6 \ M^{-2}s^{-1} .$$ (reaction 4)

Concentrations of nitric oxide produced in vivo (~100 nM) (SHIBUKI 1990) are 10 000-fold lower than the local concentrations at the nozzle of a gas cylinder. The rate of reaction between nitric oxide and oxygen at their respective physiological concentrations would, therefore, be 100 000 000 times slower. Calculations based solely on reaction with oxygen reveal that 0.7 h would be required for 100 nM nitric oxide to decrease to 50 nM (i.e., first half-life), 1.4 h to decrease from 50 nM to 25 nM and so on. Clearly, the reaction of nitric oxide with molecular oxygen cannot account for its short (1–10 s) biological half-life. The rate constant for the reaction of nitric oxide with superoxide is 6.7×10^9 $M^{-1}s^{-1}$ (HUIE and PADMAJA 1993) and, since it is bimolecular, the rate is not dependent on the square of nitric oxide concentration (reaction 5)

$$\cdot NO + O_2^{\cdot -} \rightarrow ONOO^- .$$ (reaction 5)

The fact that SOD can potentiate or prolong the biological effects of nitric oxide provides strong evidence that reaction with superoxide is a major pathway for termination of nitric oxide.

7 Formation, Decomposition, and Reactivity of Peroxynitrite

For in vitro use, peroxynitrite is readily synthesized from hydrogen peroxide and nitrous acid; the reaction mixture is rapidly quenching with sodium hydroxide to yield peroxynitrite anion (REED et al. 1974). Peroxynitrite is stable in alkaline solutions for several weeks. Dilution of the alkaline stock solution into pH 7.4 potassium phosphate results in protonation to give peroxynitrous acid (pK_a 6.8) which rapidly decomposes to yield nitrate; in the presence of other target molecules, multiple oxidation products are seen (see below). Peroxynitrite formation in vivo is almost certainly due to the reaction of nitric oxide with superoxide. This rate of formation is near the diffusion limit, which means that virtually every collision between molecules of nitric oxide and superoxide will result in a reaction to produce peroxynitrite.

At pH 7.4 and 37 °C the half-life of peroxynitrite is 1–2 s (BECKMAN et al. 1990). Peroxynitrous acid can decompose via three major pathways. Reaction of peroxynitrite with target molecules results in products characteristic of both nitrogen dioxide and hydroxyl radical (reaction 6). However, kinetic data and thermodynamic calculations indicate that no free hydroxyl radical is produced (KOPPENOL et al. 1992). We have proposed that peroxynitrous acid reacts as a vibrationally activated complex rather than via homolytic fission of peroxynitrous acid as depicted in reaction 6 (CROW et al. 1994)

$$ONOO^- + H^+ \rightleftarrows ONOOH \rightarrow \text{"} \cdot NO_2 + \cdot OH \text{"} . \qquad \text{(reaction 6)}$$

Peroxynitrite will directly nitrate phenolic compounds (HALFPENNY and ROBINSON 1952; ISCHIROPOULOS et al. 1992b; BECKMAN et al. 1992). Nitration is catalyzed by transition metals and by Cu,Zn-SOD and is thought to occur via formation of a reactive nitronium ion (reaction 7) (ISCHIROPOULOS et al. 1992b; SMITH et al. 1992). The primary consequence of this reaction in vivo is the modification of tyrosine residues in proteins, which we have detected using an antibody raised to peroxynitrite-treated keyhole limpet hemocyann (KLH, see below)

$$ONOO^- + Metal^{2+} \rightarrow NO_2^+ ... O^- —Metal^{1+} \rightarrow$$
$$protein —tyr—NO_2 + HO\text{-}Metal^{1+} + Metal^{2+} + H_2O . \qquad \text{(reaction 7)}$$

In the absence of target molecules, the primary decomposition product of peroxynitrite at neutral pH is nitrate (reaction 8)

$$ONOO^- + H^+ \rightleftarrows ONOOH \rightarrow HNO_3 \rightleftarrows NO_3^- + H^+ . \qquad \text{(reaction 8)}$$

Reactions with target molecules compete with this relatively innocuous route of decomposition. Thus the variety and yield of products from peroxynitrite will vary depending on the relative abundance of specific targets and the presence or absence of transition metals or metalloproteins.

8 Peroxynitrite: The Smart Bomb of Biological Oxidants

Transition metals are not required for either the formation or the decomposition of peroxynitrite although metals like iron and copper will catalyze nitration reactions. Decomposition of one molecule of peroxynitrous acid yields the equivalent of two potent one-electron oxidants each capable of participating in a variety of reactions. Peroxynitrous acid can nitrate phenolic rings (HALFPENNY and ROBINSON 1952; ISCHIROPOULOS et al. 1992b; BECKMAN et al. 1992), hydroxylate aromatic rings (HALFPENNY and ROBINSON 1952; HOGG et al. 1992), and oxidize lipids (RADI et al. 1991a), proteins (MORENO and PRYOR 1992), and DNA (KING et al. 1992). Under physiological conditions, peroxynitrite is sufficiently stable to diffuse several cell diameters to reach critical cellular targets before becoming protonated and decomposing. We have also seen that it can cross lipid membranes, probably in the protonated (acid) form (J. Chen and J.S. Beckman, unpublished observations). While most oxidative species are derived from peroxynitrous acid, direct reactions of the otherwise stable peroxynitrite anion occur with nucleophiles like thiols (RADI et al. 1991b) which effectively targets its oxidative capabilities. Thus, peroxynitrite may be considered the "smart bomb" of biological oxidants in that it is formed from two weak oxidants in a metal-independent manner, it can deliver multiple reactivites (including hydroxyl radical-like reactivity) over a relatively long distance, and the anionic form is stable except for its reactivity toward thiols.

9 The Dark Side of Superoxide Dismutase—Catalysis of Nitration by Peroxynitrite

Nitration of phenolic compounds by peroxynitrite is readily catalyzed by Cu, Zn-SOD (ISCHIROPOULOS et al. 1992b). Removal of the copper atom from the active site abolishes both nitrating and dismutase activities and activities are restored by copper replacement (SMITH et al. 1992). Catalysis of nitration by SOD is faster and more efficient than with either free copper or EDDA-chelated copper, suggesting that the substrate binding region of SOD recognizes peroxynitrite. This seems reasonable based on the structural similarity between peroxynitrite and super-oxide and fact that both are negatively charged. The ability of SOD to enchance nitration is modest in simple in vitro systems. However, in complex media such as brain homogenates or plasma, where competing side reactions would tend to decrease total nitration, the enhancement by SOD is pronounced and consistent with a direct reaction between peroxynitrite and SOD. Again, this difference suggests that the electrostatic field of SOD, which serves to attract the negatively charged superoxide anion, may also attract peroxynitrite anion.

In the absence of suitable targets compounds (e.g., low molecular weight phenolics or tyrosine residues in proteins), bovine Cu,Zn SOD will catalyze nitration of its own single tyrosine residue. The distance between the catalytic site and the affected tyrosine residue suggests that nitration is inter- rather than intra-molecular. Nitration renders the protein yellow at neutral pH but has no effect on dismutating or subsequent nitrating activity (SMITH et al. 1992).

10 Biological Relevance: Evidence for Peroxynitrite Formation In Vivo

Superoxide dismutase prolongs the biological half-life of either endogenous nitric oxide or authentic nitric oxide added to a tissue bath (IGNARRO et al. 1987, 1989; KELM and SCHRADER 1990; BATES et al. 1991; CHEN and LEE 1993). The most reasonable interpretation of this prolongation is that SOD scavenges superoxide, thereby preventing its reaction with nitric oxide. The transient, intermediate form of SOD, Cu^{+1}-SOD, has been reported to directly reduce nitric oxide to nitroxyl anion under anaerobic conditions (MURPHY and SIES 1991). However, it is not clear whether this would occur to any extent under physiological (aerobic) conditions in which the sole source of reductant for SOD is superoxide itself. Provided that no direct reaction occurs between nitric oxide and SOD, the ability of SOD to prolong the half-life of endogenous nitric oxide is consistent with ongoing formation of peroxynitrite even at low physiological concentrations of superoxide and nitric oxide.

Using nitration of a more water-soluble tyrosine analogue (p-hydroxyphenyl-acetic acid) as an index, we have shown that virtually all of the nitric oxide produced by activated macrophages is converted to peroxynitrite (ISCHIROPOULOS et al. 1992a). Peroxynitrite production by inflammatory cells may represent a programmed killing mechanism whereby two relatively non-toxic radical species are simultaneously and deliberately produced to form a potent oxidant directed against an invading microbe. Peroxynitrite formed outside the inflammatory cell would set up a concentration gradient favoring diffusion away from (and thereby protection of) the producing cell and toward the invader. The presence of SOD inside the microbial cell would offer no protection against preformed peroxynitrite and might actually serve to mediate injury by catalyzing nitration of microbial proteins.

Under normal physiological conditions the concentration of SOD is 100- to 1000-fold higher than nitric oxide and the rate constant for SOD's reaction with superoxide is only three fold lower (KLUG et al. 1972) than the rate constant for nitric oxide and superoxide (HUIE and PADMAJA 1993). Therefore, SOD effectively outcompetes nitric oxide for the available superoxide by approximately 30-fold and thereby provides the primary defense against host cell injury by limiting peroxynitrite formation. However, in inflammatory conditions (e.g., sepsis), in which nitric oxide production is increased by induction of a constitutively active isoenzyme of nitric oxide synthase, peroxynitrite formation would predominate and the role SOD may shift from prevention to promotion of injury by catalyzing nitration. This could account for the exacerbation of injury following exogenous administration of high doses of SOD in the same model systems in which lower doses of SOD were protective (PARKS et al. 1982; MATSUDA et al. 1991; OMAR et al. 1990).

Peroxynitrite readily nitrates tyrosine present as residues in proteins or as the free amino acid. SOD-catalyzed nitration by peroxynitrite occurs at a rate of 10^5 $M^{-1}s^{-1}$ (BECKMAN et al. 1992). Nitrogen dioxide will nitrate tyrosine at a similar rate (3.2×10^5 $M^{-1}s^{-1}$) (PRUTZ et al. 1985), but relatively high concentrations are required and the overall reaction efficiency is at least 100-fold lower than for peroxynitrite (Crow and Beckman, unpublished observations). Nitrotyrosine has been measured in human urine (OHSHIMA et al. 1990, 1991), indicating that it is stable as the free amino acid and suggesting that the nitro group is not subject to significant metabolic transformation. Thus, nitration of the phenolic ring of tryosine residues in proteins is a stable covalent modification which appears to be a specific biological marker of peroxynitrite.

We have raised an antibody in rabbits that recognizes a number of proteins treated with peroxynitrite in vitro at a titer of 1/256,000. The antibodies do not cross-react with native albumin, hemoglobin, SOD, histone, lysozyme, lung surfactant proteins, rat brain homogenates or human CSF, but recognize such proteins after treatment with peroxynitrite in both ELISA assays or western blots. The presence of nitrotyrosine has been confirmed by amino acid analysis of hydrolyzed proteins (Crow and Beckman, unpublished). Preincubation of the antibody with 10 mM nitrotyrosine blocks subsequent binding to nitrated proteins whereas aminotyrosine and phosphotyrosine do not. Treatment of nitrated proteins or tissue sections with dithionite, which reduces nitrotyrosine to aminotyrosine, (Fig. 1) also blocks binding.

We present immunohistochemical evidence (using this antibody) that lung tissue from a patient with adult respiratory distress syndrome (ARDS) (Figs. 2A, 3A) and a lesion from an atherosclerotic human artery show discrete areas of heavy staining for nitrated proteins (Fig. 2B). In lung, staining appears to be most intense in regions immediately surrounding inflammatory cells but is not restricted to those areas; note that erythrocytes immediately adjacent to inflammatory cells do not stain. In the atherosclerotic vessel, the endothelial lining stains heavily as does the plaque itself, where evidence of inflammatory cell infiltration is seen. Staining of lung is blocked by preincubation of the antibody with 10 mM nitrotyrosine (Fig. 3B) or by pretreatment of the tissue with dithionite (not shown). Using preimmune sera, a pattern of very light, homogeneous staining was seen which is essentially identical to that seen when the primary antibody is preincubated with nitrotyrosine (Fig. 3B).

The immunohistochemical results strongly indicate enhanced peroxynitrite formation in inflammatory conditions. We have shown previously that activated macrophages produce relatively large amounts of peroxynitrite under conditions of simultaneous superoxide and nitric oxide production (ISCHIROPOULOS et al. 1992a). Thus, in the case of lung and atheroma macrophages and neutrophils are a likely source of peroxynitrite.

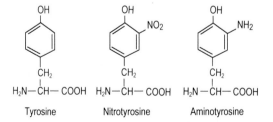

Fig. 1. Structures of the naturally occurring amino acid tyrosine and its reaction product with peroxynitrite, 3-nitrotyrosine. Nitrotyrosine, either as the free amino acid or within a protein, can be chemically reduced with dithionite (sodium hydrosulfite) to 3-aminotyrosine. Under alkaline conditions, nitrotyrosine absorbs at 420 nm whereas aminotyrosine does not. Thus, reduction can be monitored spectrally at 420 nm. This procedure is used to tentatively identify and quantify nitrotyrosine in various assays

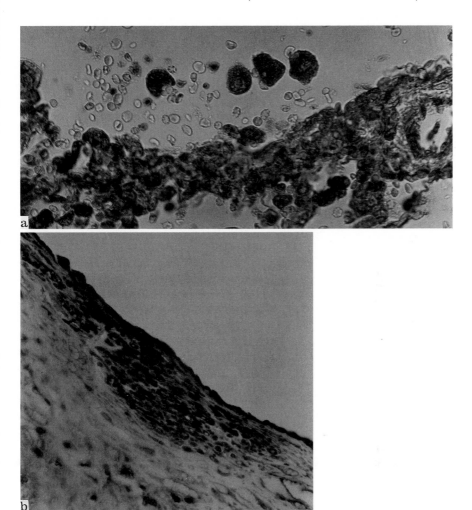

Fig. 2. a Lung tissue from a human patient with ARDS. **b** Atheromatous plaque in large vessel from human patient. Fixed tissues sections were treated with rabbit polyclonal antibody raised to peroxynitrite-treated KLH (1:1000), washed, treated with anti-rabbit globulin conjugated to biotin, washed again, and treated with avidin-conjugated horse-radish peroxidase. Peroxidase substrate was added and allowed to react. The section was then washed and counterstained with hematoxylin and eosin

11 Possible Mechanisms of Peroxynitrite-Mediated Injury

Nitration of proteins by peroxynitrite may be injurious via multiple mechanisms; alteration of tyrosine phosphorylation, altered protein function, increased protein turnover due to enhanced proteolysis, and initiation of autoimmune reactions

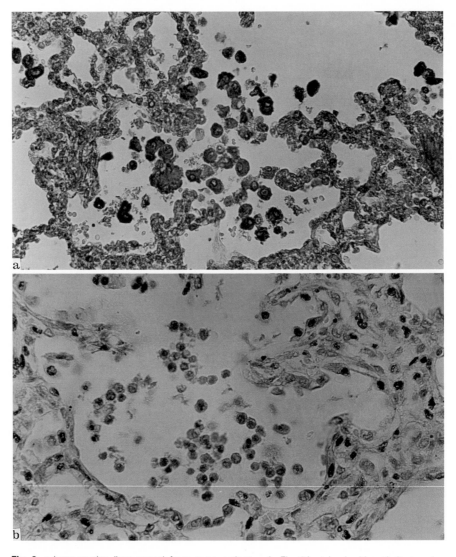

Fig. 3. a Lung section (low power) from same patient as in Fig. 2A stained with anti-nitrotyrosine antibody as in Fig. 2. **b** Same as **a** except that primary antibody was preincubated with 10 m*M* nitrotyrosine

to antigenic haptens created by protein nitration. Nitration can alter protein conformation by decreasing the hydrophobicity of tyrosine and/or by introducing a negative charge into a previously charge-neutral region of the protein. The ability of nitration of mimic phosphorylation is currently under investigation; however, the ability of nitration to inhibit tyrosine phosphorylation has been demonstrated (MARTIN et al. 1990).

Tetranitromethane has been used for many years by biochemists to nitrate tyrosines in isolated proteins to investigate their role in protein or enzyme function. Most enzymes do not depend on tyrosines for catalytic activity and many tyrosines can be nitrated without dramatically affecting activity. However, tetranitromethane inactivates the C1q binding capacity of human IgG (McCALL and EASTERBROOK-SMITH 1989), abolishes the inhibitory activity of α-1-proteinase inhibitor toward elastase (MORENO and PRYOR 1992), inhibits the activity of cytochrome P450 (JANIG et al. 1987, 1988), a-thrombin (SCULLY et al. 1992), and mitochondrial ATPase (WU and FISHER 1982; GUERRIERI et al. 1984) and inhibits the binding of human high density lipoprotein to liver plasma membranes (DARLEY-USMAR et al. 1992; GRAHM et al. 1993). Even if the activity of a given protein were unaffected by nitration, it could alter protein conformation sufficiently to "tag" it for proteolysis or cause it to be recognized as foreign by the immune system.

The technique of chemically linking dinitrophenol (which resembles nitrotyrosine) to proteins to make them more antigenic has been used for many years. This fact, together with our own experience of making an antibody to nitrated KLH protein, suggests that proteins nitrated in vivo may be highly antigenic and serve as the initiating stimulus for an autoimmune response.

12 Nitric Oxide—The Molecular Chameleon

The availability and widespread use of potent inhibitors of nitric oxide synthase has firmly established the fundamental role of nitric oxide in a wide range of physiological processes. Yet, the ability of these inhibitors to reduce or prevent injury in conditions normally associated with increased nitric oxide production (e.g., sepsis, ischemia/reperfusion) indicates that nitric oxide also has significant toxicities associated with its production. The relative lack of direct toxicity in most systems suggests that some reaction product of nitric oxide is responsible. The ability of nitric oxide to react instantaneously with superoxide, coupled with the ability of the reaction product, peroxynitrite, to permanently modify a variety of biological molecules, provides a plausible toxicological explanation and is totally consistent with the ability of SOD to prolong the biological half-life of nitric oxide and to protect against injury in response to oxidative stress. Indeed, the footprint of peroxynitrite, nitrotyrosine, is revealed in pathologic tissues by immunohistological techniques and confirmed be chemical analysis.

A current debate centers around whether the predominant effect of nitric oxide is to cause injury or to prevent it. This is particularly germane given the clinical potential of inhaled nitric oxide as therapy for pulmonary edema and possibly ARDS. This question can only be approached by considering specific conditions in which the chemistry of nitric oxide may provide some clues.

Concentrations of 10–100 nM nitric oxide are produced in vivo for signal transduction; this obviously poses no real danger to humans since we produce it continuously for 70–80 years or longer. At these levels SOD would out-compete nitric oxide for superoxide by 340-fold (based on relative concentration times reaction rate), thereby limiting peroxynitrite formation. At higher nitric oxide concentrations such as seen in ischemic brain (2–4 µM) and in the vicinity of macrophages (1–10 µM) (Ischiropoulos et al. 1992a), the ability of SOD to out-compete nitric oxide for superoxide would be greatly diminished. However, tipping the balance in favor of nitric oxide toxicity almost certainly involves other mechanisms. For example, in studies involving ischemia/reperfusion, the ability of nitric oxide to dilate vessels and thereby increase colateral bloodflow may more than offset any toxicity related to nitric oxide overproduction per se. Within the vasculature, nitric oxide could reduce injury by inhibiting both neutrophil adhesion (Kubes et al. 1991) and platelet aggregation (Macdonald et al. 1988).

Other factors which may contribute to the observed toxicity/protection by nitric oxide relate to its site of production. The constitutive nitric oxide synthase in endothelial cells appears to be bound to the plasma membrane (Pollock et al. 1991). The hydrophobic character of nitric oxide would favor its partitioning into the lipid membrane theraby preventing its reaction with the charged, hydrophilic superoxide molecule. Nitric oxide within the lipid bilayer would be positioned to react with lipid peroxyl radicals (rate constant = $3 \times 10^9 \, M^{-1}s^{-1}$) (Padmaja and Huie 1993). Formation of a stable organic nitrate (R-NO_3) could effectively terminate lipid peroxidation and thereby be protective.

We have recently investigated the reaction between nitric oxide and peroxynitrite. The products of this reaction have not been characterized, but the reaction is quite rapid and appears to destroy nitric oxide while enhancing the nitration yield from peroxynitrite (Crow and Beckman, manuscript in preparation).

It has been suggested that since peroxynitrite decomposes to innocuous nitrate, the reaction of nitric oxide with superoxide to give peroxynitrite represents a detoxification mechanism. Two crucial points must be considered. (1) Peroxynitrite is a much stronger oxidant than is nitric oxide and is more damaging in every system which has been examined; thus, the idea of forming a potent oxidant to detoxify a weaker oxidant is unsound. (2) The products of peroxynitrite reactions are dependent on the targets present. Oxidative, hydroxylating, and nitrating reactions of peroxynitrite with biological molecules compete with the decomposition pathway (to give nitrate); the primary determinants of product distribution are the relative concentrations of the reactants and the relative rate constants for the respective reactions. In a simple system in which the total concentration of all target molecules is extremely low, the predominant product of peroxynitrite would be nitrate simply because there are no other possible reaction pathways. This is analogous to comparing the results of an explosion which occurred in a crowded room with the same explosion in a empty room. Clearly, the total target concentration would be high inside or on the surface of a cell, where nitric oxide is most likely to encounter superoxide and form peroxynitrite. Thus, it is imperative that toxicity arguments be based on realistic physiological milieu and known reactivities of the species involved.

References

Bates JN, Harrison DG, Myers RR, Minor RL (1991) EDRF: nitrosylated compound or authentic nitric oxide. Basic Res Cardiol 86 [Suppl 2]: 17–26

Baum RM (1984) Superoxide theory of oxygen toxicity is center of heated debate. Chem. Engin. News 9: 20–28

Beckman JS, Beckman TW, Chen J, Marshall PA, Freeman BA (1990) Apparent hydroxyl radical production by peroxynitrite: immplications for endothelial injury from nitric oxide and superoxide. Proc Natl Acad Sci USA 87: 1620–1624

Beckman JS, Ischiropoulos H, Zhu L, van der Woerd M, Smith C, Chen J, Harrison J, Martin JC, Tsai M (1992) Kinetics of superoxide dismutase- and iron-catalyzed nitration of phenolics by peroxynitrite. Arch Biochem Biophys 298: 438–445

Berrazue JR, Lopez-Jaramillo P, Moncada S (1990) Nitric oxide: from endogenous vasodilator to biologic mediator (Review in Spanish). Rev Esp Cardiol 43: 421–431

Bielski BHJ (1978) Re-evaluation of the spectral and kinetic properties of HO_2 and O_2 free radicals. Photochem Photobiol 28: 645–649

Biemond P, van Eijk HG, Swaak AJG, Koster JF (1984) Iron mobilization from ferritin by superoxide derived from phagocytosing polymorphonuclear leucocytes. Possible mechanism in inflammation diseases. J Clin Invest 73: 1576–1579

Cazevieille C, Muller A, Meynier F, Bonne C (1993) Superoxide and nitric oxide cooperation in hypoxia/ reoxygenation-induced neuron injury. Free Radic Biol Med 14: 389–395

Chen FY, Lee TJ (1993) Role of nitric oxide in neurogenic vasodilation of porcine cerebral artery. Pharmacol Exp Ther 265: 339–345

Crow JP, Spruell C, Chen J, Gunn C, Ischiropoulos H, Tsai M, Smith CD, Radi R, Koenol WH, Beckman JS (1994) On the pH-dependent yield of hydroxyl radical products from peroxynitrite. Free radical Biol Med (in press)

Darley-Usmar VM, Hogg N, O'Leary VJ, Wilson MT, Moncada S (1992) The simultaneous generation of superoxide and nitric oxide can initiate lipid peroxidation in human low density lipoprotein. Free Radic Res Commun 17: 9–20

Dawson VL, Dawson TM, Bartley DA, Uhl GR, Snyder SH (1993) Mechanisms of nitric oxide-mediated netrotoxicity in primary brain cultures. J Neurosci 13: 2651–2661

Flaherty JT, Weisfeldt ML (1988) Reperfusion injury. Free Radic Biol Med 5: 409-419

Ford PC, Wink DA, Stanbury DM (1993) Autoxidation kinetics of aqueous nitric oxide (review). FEBS Lett 326: 1–3

Gardner PR, Fridovich I (1991a) superoxide sensitivity of the *Escherichia coli* aconitase. J Biol Chem 266: 19328–19333

Gardner PR, Fridovich I (1991b) Superoxide sensitivity of the *Escherichia coli* 6-phosphogluconate dehydratase. J Biol Chem 266: 1478–1483

Gardner PR, Fridovich I (1992) Inactivation-reactivation of aconitase in *Escherichia coli*. A sensitive measure of superoxide radical. J Biol Chem 267: 8757–8763

Garside P, Hutton AK, Severn A, Liew FY, Mowat AM (1992) Nitric oxide mediates intestinal pathology in graft-vs.-host disease. Eur J Immunol 22: 2141–2145

Graham A, Hogg N, Kalyanaraman B, O'Leary V, Darley-Usmar V, Moncada S (1993) Peroxynitrite modification of low-density lipoprotein leads to recognition by the macrophage scavenger receptor. FEBS Lett 330: 181–185

Green SJ, Mellouk S, Hoffman SL, Meltzer MS, Nacy CA (1990a) Cellular mechanisms of nonspecific immunity to intracellular infection: cytokine-induced synthesis of toxic nitrogen oxides from L-arginine by macrophages and hepatocytes. Immunol Lett 25: 15–19

Green SJ, Meltzer MS, Hibbs JB, Jr Nacy CA (1990b) Activated macrophages destory intracellular *Leishmania major* amastigotes by an L-arginine-dependent killing mechanism. J Immunol 144: 278–283

Guerrieri F, Yagi A, Yagi T, Papa S (1984) On the mechanism of H^+ translocation by mitochondrial H+ -ATPase. Studies with chemical modifier of tyrosine residues. J Bioenerg Biomembr 16: 251–262

Haber F, Weiss J (1934) The catalytic decomposition of hydrogen peroxide by iron salts. Proc R Soc 147: 332–351

Halfpenny E, Robinson PL (1952) The nitration and hydroxylation of aromatic compounds by pernitrous acid. J Chem Soc 939–946

Hoffman RA, Langrehr JM, Simmons RL (1992) The role of inducible nitric oxide synthetase during graft-versus-host disease. Transplant Proc 24: 2856

Hogg N, Darley-Usmar VM, Wilson MT, Moncada S (1992) Production of hydroxyl radicals from the simultaneous generation of superoxide and nitric oxide. Biochem J 281: 419–424

Huie RE, Padmaja S (1993) The reaction of NO with superoxide. Free Radic Res Commun 18: 195–199

Hutchinson PJ, Palmer RM, Moncada S (1987) Comparative pharmacology of EDRF and nitric oxide on vascular strips. Eur J Pharmacol 141: 445–451

Ignarro LJ, Buga GM, Wood KS, Byrns RE, Chaudhuri G (1987) Endothelium-derived relaxing factor produced and released from artery and vein is nitric oxide. Proc Natl Acad Sci USA 84: 9265–9269

Ignarro LJ, Gold ME, Buga GM, Byrns RE, Wood KS, Chaudhuri G, Frank G (1989) Basic polyamino acids rich in arginine, lysine, or ornithine cause both enhancement of and refractoriness to formation of endothelium-derived nitric oxide in pulmonary artery and vein. Circ Res 64: 315–329

Ischiropoulos H, Zhu L, Beckman JS (1992a) Peroxynitrite formation from macrophage-derived nitric oxide. Arch Bioch Biophys 298: 446–451

Ischiropoulos H, Zhu L, Chen J, Tsai M, Martin JC, Smith CD, Beckman JS (1992b) Peroxynitrite-mediated tyrosine nitration catalyzed by superoxide dismutase. Arch Biochem Biophys 298: 431–437

Jaeschke H, Schini VB, Farhood A (1992) Role of nitric oxide in the oxidant stress during ischemia/reperfusion injury of the liver. Life Sci 50: 1797–1804

Janig GR, Kraft R, Blanck J, Ristau O, Rabe H, Ruckpaul K (1987) Chemical modification of cytochrome P-250 LM4. Identification of functionally linked tyrosine residues. Biochim Biophys Acta 91: 512–523

Janig GR, Kraft R, Rabe H, Makower A, Ruckpaul K (1988) Comparative studies on the accessibility and functional importance of tyrosine residues in cytochrome P-450 isozymes. Biomed Biochim Acta 47: 565–579

Kelm M, Schrader J (1990) Control of coronary vascular tone by nitric oxide. Circ Res 66: 1561–1575

King PA, Anderson VE, Edwards JO, Gustafson G, Plumb RC, Suggs JW (1992) A stable solid that generates hydroxyl radical upon dissolution in aqueous solution: Reaction with proteins and nucleic acid. J Am Chem Soc 114: 5430–5432

Klug D, Rabani J, Fridovich I (1972) A direct demonstration of the catalytic action of superoxide dismutase through the issue of pulse radiolysis. J Biol Chem 247: 4839–4842

Koppenol WH, Moreno JJ, Pryor WA, Ischiropoulos H, Beckman JS (1992) Peroxynitrite, a cloaked oxidant formed by nitric oxide and superoxide. Chem Res Toxicol 5: 834–842

Kroncke KD, Kolb-Bachofen V, Berschick B, Burkart V, Kolb H (1991) Activated macrophages kill pancreatic syngeneic islet cells via arginine-dependent nitric oxide generation. Biochem Biophys Res Commun 175: 752–758

Kubes P, Suzuki M, Granger DN (1991) Nitric oxide: an endogenous modulator of leukocyte adhesion. Proc Nat Acad Sci USA 88: 4651–4655

Kwon NS, Stuehr DJ, Nathan CF (1991) Inhibition of tumor cell ribonucleotide reductase by macrophage-derived nitric oxide. J Exp Med 174: 761–767

Langrehr JM, Murase N, Markus PM, Cai X, Neuhaus P, Schraut W, Simmons RL, Hoffman RA (1992) Nitric oxide production in host-versus-graft and graft-versus-host reactions in the rat. J Clin Invest 90: 679–683

Lepoivre M, Fieschi F, Coves J, Thelander L, Fontecave M (1991) Inactivation of ribonucleotide reductase by nitric oxide. Biochem Biophys Res Commun 179: 442–448

Macdonald PS, Read MA, Dusting GJ (1988) Synergistic inhibition of platelet aggregation by endothelium-derived relaxing factor and prostacyclin. Thromb Res 49: 437–449

Martin BL, Wu D, Jakes S, Graves DJ (1990) Chemical influences on the specificity of tyrosine phosphorylation. J Biol Chem 265: 7108–7111

Masini E, Bianchi S, Mugai L, Gambassi F, Lupini M, Pistelli A, Mannaioni PF (1991) The effect of nitric oxide generators on ischemia reperfusion injury and histamine release in isolated perfused guinea-pig heart. Agents Actions 33: 53–56

Matsuda M, Fujiwara H, Kawamura A, Ishida M, Takemura G, Kida M, Uegaito T, Fujiwara Y, Fujiwara T, Kawai C (1991) Failure to reduce infarct size by intracoronary infusion of recombinant human superoxide dismutase at reperfusion in the procine heart—immunohistochemical and histological analysis. J Mol Cell Cardiol 23: 1287–1296

McCall MN, Easterbrook-Smith SB (1989) Comparison of the role of tyrosine residues in human IgG and rabbit IgG in binding of complement subcomponent C1q. Biochem J 257: 845–851

McCord JM, Fridovich I (1969) Superoxide dismutase and enzymic function for erythrocuperein (hemocuprein). J Biol Chem 244: 6049–6055

Moreno JJ, Pryor WA (1992) Inactivation of alpha 1-proteinase inhibitor by peroxynitrate. Chem Res Toxicol 5: 425–431

Mulligan MS, Hevel JM, Marletta MA, Ward PA (1991) Tissue injury caused by deposition of immune complexes is L-arginine dependent. Proc Natl Acad Sci USA 88: 6338–6342

Murphy ME, Sies H (1991) Reversible conversion of nitroxyl anion to nitric oxide by superoxide dismutase. Proc Natl Acad Sci USA 88: 10860–10864

Nguyen T, Brunson D, Crespi CL, Penman W, Wishnok JS, Tannebaum, SR (1992) DNA damage and mutation in human cells exposed to nitric oxide in vitro. Proc Nat Acad Sci USA 89: 3030–3034

Ohshima H, Friesen M, Brouet I, Bartsch H (1990) Nitrotyrosine as a new marker for endogenous nitrosation and nitration of proteins. Food Chem Toxicol 28: 647–652

Ohshima H, Brouet I, Friesen M, Bartsch H (1991) Nitrotyrosine as a new marker for endogenous nitrosation and nitration. IARC Scientific Publications, Lyon, 443–448

Omar BA, Gad NM, Jordan MC, Striplin SP, Russell WJ, Downey JM, McCord JM (1990) Cardioprotection by Cu,Zn-superoxide dismutase is lost at high doses in the reoxygenated heart. Free Radic Biol Med 9: 465–471

Padmaja S, Huie RE (1993) The reaction of nitric oxide with organic peroxyl radicals. Biochem Biophys Res Commun 195: 539–544

Parks DA, Bulkley GB, Granger DN, Hamilton SR, McCord JM (1982) Ischemic injury in the cat small intestine: role of superoxide radicals. Gastroenterology 82: 9–15

Pauling L (1979) The discovery of the superoxide radical. Trends Biochem Sci 4: N270–N271

Pollock JS, Forstermann U, Mitchell JA, Warner TD, Schmidt HHHW, Nakane M, Murad F (1991) Purification and characterization of particulate endothelium-derived relaxing factor synthase from cultured and native bovine aortic endothelial cells. Proc Natl Acad Sci USA 88: 10480–10484

Prutz WA, Monig H, Butlker J, Land EJ (1985) Reactions of nitrogen dioxide in aqueous model systems: oxidation of tyrosine units in peptides and proteins. Arch ;Biochem Biophys 243: 125–134

Radi R, Beckman JS, Bush KM, Freeman BA (1991a) Peroxynitrite-induced membrane lipid peroxidation: the cytotoxic potential of superoxide and nitric oxide. Arch Biochem Biophys 288: 481–487

Radi R, Beckman JS, Bush KM, Freeman BA (1991b) Peroxynitrite oxidation of sulfhydryls. The cytotoxic potential of superoxide and nitirc oxide. J Biol Chem 266: 4244–4250

Reed JW, Ho HH, Jolly WL (1974) Chemical synthesis with a quenched flow reactor. Hydroxytrihydroborate and peroxynitrite. JACS 96: 1248–1249

Rush JD, Koppenol WH (1989) Reactive intermediates formed by the interaction of hydrogen peroxide and ferrous complexes. In: Beaumont P, Deeble D, Parsons B, Rice-Evans C (eds) Free radicals, metal ions and biopolymers. Richelieu, London, pp 33–44

Saran M, Michel C, Bors W (1990) Reaction of NO with O_2.. implications for the action of endothelium-derived relaxing factor (EDRF). Free Radic Res Commun 10: 221–226

Sawyer DT, Valentine JS (1981) How super is superoxide? Accts Chem Res 14: 393–400

Scully MF, Kakkar VV, Goodwin CA (1992) Non-specific influence of chemical modification upon the properties of antithrombin III: modification of carboxyl groups. Thromb Res 67: 447–456

Shibuki K (1990) An electrochemical microprobe for detecting nitric oxide release in brain tissue. Neurosci Res 9: 69–76

Smith CD, Carson M, van der Woerd M, Chen J, Ischiropoulos H, Beckman JS (1992) Crystal structure of peroxynitrite-modified bovine Cu,Zn-superoxide dismutase. Arch Biochem Biophys 299: 350–355

Thomas EE, Morehouse LA, Aust SD (1985) Ferritin and superoxide-dependent lipid peroxidation. J Biol Chem 260: 3275–3280

Welsh N, Eizirik DL, Bendtzen K, Sandler S (1991) Interleukin-1 beta-induced nitric oxide production in isolated rat pancreatic islets requires gene transcription and may lead to inhibition of the Krebs cycle enzyme aconitase. Endocrinology 129: 3167–3173

Wink DA, Kasprzak KD, Maragos CM, Elespuru RK, Misra M, Dunams TM, Cebula TA, Koch WH, Andrews AW, Allen JS et al. (1991) DNA deaminating ability and genotoxicity of nitric oxide and its progenitors. Science 254: 1001–1003

Wu LN, Fisher RR (1982) Modification of bovine heart mitochondrial transhydrogenase with tetranitromethane. Biochim Biophys Acta 681: 388–396

Zweier JL, Rayburn BK, Flaherty JT, Weisfeldt ML. (1987) Recombinant superoxide dismutase reduces oxygen free radical concentrations in reperfused myocardium. J Clin Invest 80(6): 1728–1734

Regulation and Dysregulation of Constitutive Nitric Oxide Synthases Types I and III

H.H.H.W. SCHMIDT, H. HOFMANN, and P. OGILVIE

1 Introduction

Mammalian cells synthesize nitric oxide (NO) from L-arginine by NO synthases (NOS), members of the cytochrome P_{450} gene family. NO represents the best understood member of a novel class of redox-active transmembrane and intracellular signaling molecules that may also include CO and oxygen radicals. When formed and released in nanomolar concentrations, the main biological effect of NO is to trigger the guanosine cyclic 3',5'-monophosphate signalling cascade by activating the cGMP-generating enzyme soluble guanylyl cyclase (SCHMIDT et al. 1993).

2 Nitric Oxide Synthase Isozymes

In mammalian cells, NO is enzymatically formed from a terminal guanidino-nitrogen of L-arginine (HIBBS et al. 1987; PALMER et al. 1988; SCHMIDT et al. 1988 a,b) by a family of at least three distinct NOS isozymes (E.C. 1.14.23) (SCHMIDT et al. 1991). All described NOSs yield L-citrulline as a coproduct of this reaction.

Department of Clinical Biochemistry and Pathobiochemistry, Medical University Clinic, Versbacher Strasse 5, 97078 Würzburg, Germany

Table 1. Structural and Kinetic characteristics of nitric oxide synthase

Parameter binding site	Type I (brain)	Type II (macrophage)	Type III (endothelial)
Structure			
Native	Homodimer	Homodimer	?
Denatured, kDa	160	130	133
Calmodulin			
Binds	Reversible	Copurifies	Reversible
Essential	Yes	?	Yes
Regulation	Free Ca^{2+}	Constitutive	Free Ca^{2+}
L-Arginine			
K_m (μm)	2–7	3–32	3
V_{max}, (nmol/mg per min)	≥ 300	≤ 1600	15
Co/posttranslational modifications			
Phosphorylation	Yes	?	Yes
Myristoylation	No	No	Yes
Subcellular localization (100 000 × g)	Soluble	Soluble	Posttranslationally regulated

The original classification of NOS into NOS-I, II, and III (Table 1) was based on the physical and biochemical characteristics of the purified enzymes, i.e., subcellular location (soluble vs particulate fraction) and regulation by the free Ca^{2+} concentration (SCHMIDT et al. 1991), and has been confirmed by the recent molecular cloning and expression studies of three corresponding genes (BREDT et al. 1991; MARSDEN et al. 1992; XIE et al. 1992; NAKANE et al. 1993). Other current classifications group NOS isoforms according to their: (1) regulation of expression (constitutive and inducible NOS), (2) regulation by free Ca^{2+} (Ca^{2+}-dependent and Ca^{2+}-independent NOS; (NATHAN 1992)), or (3) primary source for purification (brain-, macrophage-, and endothelial-type NOS) (SNYDER and BREDT 1991; MARSDEN et al. 1992; NATHAN 1992; SESSA et al. 1992). However, NOS-I, for example, is not exlusively expressed in brain but also in the peripheral nervous system and in nonneuronal cells (SCHMIDT et al. 1992e; WILCOX et al. 1992), and its expression is not constitutive under all conditions (see below). Moreover, the terms "constitutive" or "Ca^{2+}-dependent" NOS do not differentiate between type I and type III, which derive from two distinct genes (see below).

3 Nitric Oxide Synthase Structure

Nitric oxide synthases are homodimers (NOS_2) (SCHMIDT et al. 1991) of subunits which range between 130 and 160 kDa (BREDT et al. 1991; JANSSENS et al. 1992; LOWENSTEIN et al. 1992; LYSONS et al. 1992; MARSDEN et al. 1992; SESSA et al. 1992; XIE et al. 1992; NAKANE et al. 1993). The COOH-terminal half of all NOS

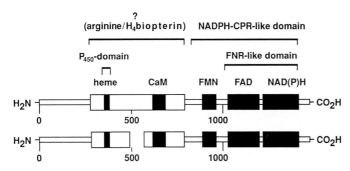

Fig. 1. Consensus sequences for binding sites and protein domains of nitric oxide synthases (NOSs). Catalytic sites (*filled boxes*) include the cytochrome P_{450}-like oxidase (heme binding site) and the P_{450} reductase (*CPR*)-like reductase domain (FMN, FAD, NADPH binding sites). Regulatory domains include the calmodulin binding site; several potential phosphorylation sites exist (not shown). The L-arginine and tetrahydrobiopterin (H_4biopterin) binding sites may locate to a conserved region in the NH_2-terminal half (*open boxes*). An in-frame deletion of this region and an NOS-I greatly reduced in activity and smaller by about 10 kDa has been described

contains bindings motifs for NADPH, FAD and FMN, identical to NADPH-cytochrome P_{450} reductase (CPR), one other flavin nucleotide reductase that contains both flavin nucleotides, FMN and FAD (Fig.1). In addition, the NOS heme moiety presents the typical CO difference spectrum of P_{450}-type enzymes. Due to its CPR-like domain NOS can, thus, be viewed as a mechanistically self-contained cytochrome P_{450}/CPR chimeric protein which combines both an oxidase and a reductase domain in a single polypeptide. All NOS bind calmodulin in a process which is either Ca^{2+}-dependent (NOS-I and III) or Ca^{2+}-independent (NOS-II) (CHO et al. 1992; SCHMIDT et al. 1992b). In the case of NOS-II, calmodulin and enzyme copurify, suggesting a native heterotetrameric structure (NOS_2/calmodulin$_2$) of this isoform. Calmodulin is the key regulator of electron flow within NOS (HEINZEL et al. 1992; KLATT et al. 1992). Additional binding sites within the NOS polypeptide must be postulated for tetrahydrobiopterin (BH_4) and L-arginine, but have not been identified. BH_4 is utilized as a cofactor by several other amino acid hydroxylases. However, no other BH_4 utilizing enzyme but NOS binds BH_4 with an affinity that permits copurification of enzyme and cofactor (up to 2 mol per dimer). This fact and some of the effects of BH_4 on NO catalysis (see below) suggest a novel BH_4 binding site in NOS. Moreover, dimerization of NOS is a prerequisite for L-arginine turnover and BH_4-dependent (STUEHR et al. 1991). The NH_2-terminal region C of NOS (Fig.2) is very likely to contain the domains for BH_4 binding, L-arginine binding and possibly subunit dimerization. Hydrodynamic characterization of NOS suggests not only a homodimeric structure, but revealed also a frictional ratio indicative of an elongated tertiary structure hindering intramolecular contact between the oxidase domain (NH_2-terminal) and the oxidase domain (COOH-terminal). A yet hypothetical head-to-tail arrangement of NOS at the C region would allow for intermolecular contact and electron flow between the oxidase domain of one monomer and the reductase domain of another monomer (Fig. 2).

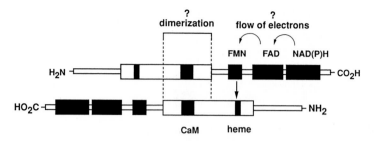

Fig. 2. Hypothetical model for nitric oxide synthase (NOS) dimerization and electron flow. Native NOS monomers have the hydrodynamic characteristics of an elongated polypeptide. Dimerization in a hypothetical head-to-tail orientation would allow for electron flow between the oxidase and reductase domains. This tertiary structure of NOS may be regulated by calmodulin and region C (*white box*), which is conserved between all NOSs

Interestingly a mutant of NOS-I mRNA bearing an in-frame deletion in region C has recently been reported in different neuroblastoma cell lines. A NOS-I truncated by about 10 kDa, but not originating from the same in-frame deletion, is expressed in embryonal rat brain and appears to have a greatly reduced specific activity.

4 Mechanisms of Nitric Oxide Synthesis

The oxidation of a terminal guanidino nitrogen of L-arginine to NO is broadly accepted as a working hypothesis to explain the mechanism of NOS catalysis (HEVEL et al. 1991; MAYER et al. 1992; SCHMIDT et al. 1992b; STUEHR and GRIFFITH 1992), whereas considerable debate concerns additional reactions NO may undergo postsynthesis, leading to storage or transport forms of this messenger molecule. Three, possibly four, cofactors (heme, FMN, FAD and H_4 biopterin) and two cosubstrates (O_2 and NADPH) participate in enzymatic NO formation. Electron flow is regulated by calmodulin binding. Dissociation of traces of OHArg from NOS were identified by HPLC and GC/MS. EPR data also suggest the formation of a N^ω-hydroxy-L-arginine (OHArg) cation radical (PRÓNAI et al. 1991). Since OHArg is utilized as an alternative substrate, it is believed to represent an intermediate of the NOS reaction. The stoichiometry of the electron transfer can fully be explained by the use of 1.5 mol NADPH per mol NO formed. It is not clear whether BH_4 is required as an additional cofactor or NADPH-dependently recycled. In all other known BH_4-utilizing enzymes, the quinoid form of dihydrobiopterin (q-BH_2) is the product of normal catalysis and is regenerated to BH_4 by a NADH-dependent q-BH_2 reductase (dihydropteridine reductase). Recyclable q-BH_2 was not detected with catalytically active NOS. However, NOS copurifies with BH_4 indicative of an inherent, NADPH-dependent biopterin reductase activity (SCHMIDT et al. 1992b). Alternatively, it was suggested that BH_4 is not metabolized during normal catalysis but acts as an allosteric activator or stabilizer of the active center (GIOVANELLI et al. 1991). The BH_4 binding site of NOS also appears to be different from that in other known BH_4-metabolizing enzymes as indicated by: (1) a uniquely small K_a value (CRAINE et al. 1972; FIRGAIRA et al. 1981), (2) high substrate specificity (KWON et al. 1989; TAYEH and MARLETTA 1989; GIOVANELLI et al. 1991) of NOS for BH_4, and (3) the resistance of NOS to inhibition even by high concentrations of methotrexate, which inhibits dihydrofolate and dihydropteridine reductases (GIOVANELLI et al. 1991).

The oxidase and reductase activities of NOS extend to other substrates (Table 2). Many of the reductions catalyzed by NOS are likely to be related to its CPR-like domain and generate an electron pressure towards the oxidase domain. As expected from the close similarity of NOS to CPR, NOS reduces cytochrome c in vitro. The mechanism of cytochrome c reduction by NOS is unclear. NOS forms superoxide anions (see above), as does CPR, but also interacts directly with cytochrome c by a highly affinity protein–protein interaction, whereas CPR does not bind to cytochrome c (KLATT et al. 1992). Furthermore, NOS also reduces cytochrome P_{450}, as indicated by the fact that, in vitro, NOS supports the hydroxylation of N-ethylmorphine by cytochrome P_{450}. This suggests that NOS may participate in similar electron transfer processes in vivo (KLATT et al. 1992) and NOSs may be viewed rather as isoforms of CPR, with L-arginine turnover being merely a side effect of their catalytic activities. The close similarity between NOS and the CPR/cytochrome P_{450} system is underscored by the fact that cytochrome P_{450} catalyzes the oxidation of OHArg to NO.

While reduction of cytochromes by NOS is strictly calmodulin-dependent, several artificial electron acceptors, e.g., nitroblue tetrazolium (NBT) and dichlorphenon-indophenol (DCPIP), are calmodulin-*independently* reduced by NOS. This suggests that NOS has: (a) a single reductase domain that is partially dependent on calmodulin binding, depending on the substrate; (b) two reductase domains, one calmodulin-dependent (oxygen, cytochromes) and one calmodulin-independent (biopterin, NBT, DCPIP). The electron acceptor NBT is converted to blue diformazan by NOS and many other dehydrogenases and reductases. Only in the case of NOS, however, is this so-called NADPH-diaphorase activity of NOS remarkably resistant to commonly used protein fixatives, thus enabling the convenient histochemical localization of NOS in paraformaldehyde-treated tissue sections (HOPE et al. 1991; SCHMIDT et al. 1992c). The utilization of NBT by NOS makes NBT also a potent noncompetitive NOS inhibitor, presumably by competing with molecular oxygen for reducing equivalents (SCHMIDT et al. 1992a,b) or by reacting with intermediate superoxide radicals. The aforementioned putative biopterin reductase activity of NOS may be closely related to the NADPH-diaphorase domain of NOS (HOPE et al. 1991).

Table 2. Reactions catalyzed by nitric oxide synthase

Type	Substrate	Cosubstrate	Regulation	Product	Reaction name
Oxidase	O_2	NADPH	Calmodulin	Superoxide, peroxide	NADPH Oxidase
Oxidoreductase	L-Arginine	NADPH, O_2, (BH_4)	Calmodulin	Nitric oxide	Nitric Oxide Synthase
Reductase C	Cyt P_{450}	NADPH, O_2	Calmodulin	Reduced Cyt P-450	CPR
Reductase C	Cyt c	NADPH (direct)	Calmodulin	Reduced Cyt c	Cyt c reductase
Reductase C, I	NBT	NADPH	Partially by calmodulin	Formazan	NADPH diaphorase
Reductase I	DCPIP	NADPH	Independent of calmodulin		

Cyt, Cytochrome; CPR, Cyt P_{450} reductase; BH_4, tetrahydrobiopterin; NBT, nitroblue tetrazolium; DCPIP, dichlorphenol-indophenol.

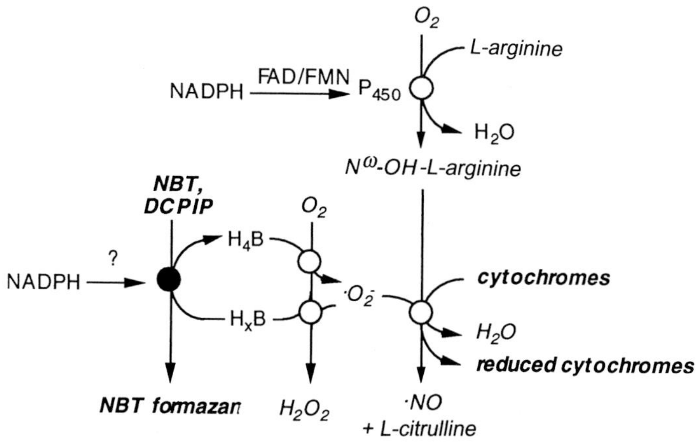

Fig. 3. Proposed reaction mechanisms for nitric oxide (*NO*) synthase. Molecular oxygen is incorporated into both NO and L-citrulline. The flavin cofactors FAD and FMN and the cytochrome P_{450} domain of NOS mediate part of or the entire electron transfer from NADPH to molecular oxygen (oxidase domain). Tetrahydrobiopterin (H_4B) may be involved in this or regulate NOS allosterically. L-Arginine is hydroxylated to the intermediate N^{ω}-hydroxy-L-arg *OHArg*. In the calcium-regulated NOS-I, two reductase activities can be distinguished by their mode of regulation: one that transfers electrons independently of calmodulin (*filled circle*) to artificial acceptors nitro blue tetrazolium (*NBT*) and dichlorphenol-indophenol (*DCPIP*) and may represent a putative biopterin reductase domain of NOS, and another one that transfers electrons in a calmodulin-dependent manner (*open circles*) to cytochromes either directly or via superoxide anions

5 Regulation of Nitric Oxide Synthase Expression and Posttranslational Modification

Types I and III NOS are constitutively expressed and in a very cell-specific manner (Table 3), whereas NOS-II is not constitutive but inducible in virtually every cell after appropriate immunological activation with different cytokines or endotoxins

Table 3. Cells and tissues expressing nitric oxide synthase or nitric oxide synthase immunoreactivity

Type I	Type II	Type III
Neurons	Macrophages	Endothelium
Macula densa	Cardiomyocytes	
Bronchial epithelium	Vascular smooth muscle	
Gastric epithelium	Hepatocytes (Calmodulin-dependent)	
HIT-T15 pancreatic B cells	Intestinal epithelium	
Mast cells	Megakaryocytes	
Human skeletal muscle	Keratinocytes	
Endothelium		
Testis		
Photoreceptors		

Not identified: Polymorphonuclear-leukocytes, platelets, bone, retina, olfactory epithelium.

Table 4. Physiological and pathophysiological regulation of nitric oxide synthase expression and activity

Level	Type I	Type II	Type III
Expression			
Constitutive	Yes	No	Yes
Increased by	Inflammation	IFN-γ, LPS, IL-1,2	Shear stress
Decreased by	IFN-γ, LPS	glucocorticoids, TGF-β, IL-4, 10	TNF-α
Mutations	In-frame deletion	?	?
Posttranslational modifications			
Dimerization	Yes	Yes (cofactor-dependent)	?
Phosphorylation	Yes	?	Yes
Myristoylation	No	No	Yes
Calmodulin binding	Reversible (Ca^{2+})	Irreversible	Reversible (Ca^{2+})
Activity			
Elevated free Ca^{2+}	Essential	No effect	Essential
Endogenous MeArg	Inhibit	Inhibit	Inhibit
Autoinhibition (NO)	No/Yes	No	?
Intracellular pH	?	?	Modulates

IFN-γ, interferon-γ; LPS, lipopolysaccharide; IL, interleukin, TGF-β, transforming growth factor-β; TNF-α, tumor necrosis factor; MeArg, methyl-L-arginine; NO, nitric oxide.

(Table 4). Once expressed, NOS-II binds calmodulin irreversibly, independently of Ca^{2+} and stays maximally active irrespective of the free Ca^{2+} concentration. The induction of NOS-II by cytokines is transcriptionally based. In control cells, mRNA for NOS-II is not detectable.

Modulation of basal NOS-I and III expression may also take place. For example, the same cytokines which induce NOS-II in RAW macrophages have the reverse effect on basal expression of the constitutive type I NOS in this cell line. The mechanisms which lead to reduced expression and the pathophysiological significance of this bidirectional, immunological regulation of NOS expression are unclear (SCHMIDT et al. 1992d; FÖRSTERMANN et al. 1993). Moreover, under physiological conditions, NOS-I expression is also not constant. We were recently able to show that during brain development enzymatically competent NOS-I is only expressed with the onset of synaptogenesis. Here, expression is dramatically increased within 1–2 days and returns to a lower steady state level at the end of synaptogenesis. The peak of expression is also different in different brain regions in pre- but not postnatal brain tissue. No NOS activity but a 150 kDa NOS-I immunoreactive protein was detected. Since the antibody that was used is monospecific for type I NOS, it is likely that this protein represents an embryonal NOS-I deletion mutant (Ogilvie et al., unpublished).

Further regulation of NOS expression takes place at the co- and posttranslational level. As mentioned before, the subcellular distribution of the various NOS isoforms is different. Even the same isoform may be distributed at varying ratios between soluble and particulate cell fractions in different species (SCHMIDT et al. 1992d). However, none of the described NOSs contain an amino acid sequence suggestive of a transmembrane domain (BREDT et al. 1991; LYONS et al. 1992). Therefore, co- or posttranslational modifications such as myristoy-lation (POLLOCK et al. 1992) and phosphorylation (MICHEL et al. 1993) are very likely to regulate the subcellular localization of NOS. These modifications may also represent an additional means to regulate enzyme activity (SCHMIDT et al. 1992a). The membrane-bound NOS-III is not only myristoylated but has also the lowest specific activity of all known NOSs (POLLOCK et al. 1991). At least for the types I and II isoforms there is considerable evidence that dimerization or an essential conformational change, as detected by gel permeation chromatography, depends on saturating concentrations of BH_4 and possibly also of L-arginine and heme. NOS monomers appear to be inactive.

6 Regulation of Nitric Oxide Synthase Activity

Once expressed, binding of the Ca^{2+}-binding protein calmodulin or phos-phorylation is an established mechanism by which enzyme activity of different isoforms of NOS is regulated (Table 1). All NOSs bind calmodulin and have conserved consensus sequences for calmodulin binding. In the case of NOS-I and III, interaction with calmodulin depends on elevated intracellular free Ca^{2+} concentrations ($[Ca^{2+}]_i$). At resting $[Ca^{2+}]_i$ (< 100 nM), these NOSs are inactive. They bind calmodulin and become fully active at increased $[Ca^{2+}]_i$ (≥ 500 nM). Calmodulin antagonists, e.g., calmidazolium and trifluoperazine, block calmodulin binding and inhibit Ca^{2+}-induced NO formation. Both compounds more potently inhibit NOS-I than NOS-III, which probably reflects differences in the respective calmodulin binding domains. In various systems, transmembrane Ca^{2+} flux is initiated by the binding of a receptor agonist to its membrane receptor. Thus, receptor occupation and increased $[Ca^{2+}]_i$ can be linked to increased NO formation in cells.

Two forms of NOS-II have been isolated, one from liver (EVANS et al. 1992) and one from a macrophage cell line (HEVEL et al. 1991; STUEHR et al. 1991). At < 100 nM free Ca^{2+}, the liver NOS-II is calmodulin-free and inactive. Similar to NOS-I and III, activity increases upon reconstitution of liver NOS-II with calmodulin, but, unlike NOS-I and III, this activation is Ca^{2+}-independent. Macrophage NOS-II has calmodulin constitutively bound and is constitutively active. Three other proteins bind calmodulin also constitutively, i.e., in an apparently Ca^{2+}-independent manner: phosphorylase kinase, a cyclic nucleotide phosphodiesterase, and Bordetella pertussis adenylyl cyclase (CHO et al. 1992). Thus, calmodulin binding represents a common activation principle for all NOSs, whereas their dependency on the free intracellular Ca^{2+} concentration distinguishes them.

The predicted amino acid sequence of NOS-I contains consensus sites for phosphorylation by cAMP-dependent protein kinases (BREDT et al. 1991). Forskolin-induced increases in intracellular cAMP levels (SCHMIDT et al.1992d) and cAMP-dependent protein kinases (BRÜNE and LAPETINA 1991) do not regulate NOS, whereas protein kinase C, Ca^{2+}/calmodulin-dependent kinase (NAKANE et al. 1991) and the phosphatase inhibitor okadaic acid (SCHMIDT et al. 1992d) do. In vitro, Ca^{2+}/calmodulin-dependent protein kinase II phosphorylates NOS on both serine and threonine (NAKANE et al. 1991). Phosphorylation is Ca^{2+}- and calmodulin-dependent and results in a marked decrease of NOS activity. Thus, frequent increases in $[Ca^{2+}]_i$ and phosphorylation of NOS by Ca^{2+}/calmodulin-dependent kinase II may represent a negative feedback regulation of NOS activity. Since Ca^{2+}/calmodulin kinase II becomes Ca^{2+}-independent upon auto-phosphorylation, it is conceivable that phosphorylation and inactivation of NOS proceed even after $[Ca^{2+}]_i$ has returned to basal levels. However, these in vitro observations need to be confirmed at the in vivo level.

Recent work has also focused on autoinhibition of NOS activity. Namely, several laboratories have identified circulating endogenous N^{ω}-methyl-L-arginine (MeArg) and Me_2Arg, which accumulate during renal failure and will systematically inhibit NO biosynthesis, possibly explaining some of the immunological dysfunctions observed under this condition. Whether under other conditions tissue levels are high enough to affect NOSs remains to be established. Moreover, several laboratories have suggested that NO acts back on NOS in a negative autofeedback loop. At least for NOS-I, we were unable to observe effects on citrulline formation either by heme, a NO scavenger, or superoxide dismutase, which should increase the apparent half-life of NO and potentiate auto-inhibition (Hofmann and Schmidt, unpublished). Thus, modulation of NO biosynthesis may occur via endogenous arginine derivatives but possibly not through NO-mediated autoinhibition.

References

Bredt DS, Hwang PM, Glatt CE, Lowenstein C, Reed RR, Snyder SH (1991) Cloned and expressed nitric oxide synthase structurally resembles cytochrome P-450 reductase. Nature 351: 714–718

Brüne B, Lapetina EG, (1991) Phosphorylation of nitric oxide synthase by protein kinase. Biochem Biophys Res Commun 181: 921–926

Cho HJ, Xie Q-W, Calaycay J, Mumford RA, Swiderek KM, Lee TD, Nathan C (1992) Calmodulin is a subunit of nitric oxide synthase from macrophages. J Exp Med 176: 599–604

Craine JE, Hall ES, Kaufman S (1972) The isolation and characterization of dihydropteridine reductase from sheep liver. J Biol Chem 247: 6082–6091

Evans T, Carpenter A, Cohen J (1992) Purification of a distinct isoform of endotoxin-induced nitric oxide synthase from rat liver. Proc Natl Acad Sci USA 89: 5361–5365

Firgaira FA, Cotton RGH, Danks DM (1981) Isolation and characterization of dihydropteridine reductase from human liver. Biochem J 197: 31–43

Förstermann U, Kuk J, Nakane M, Pollock J (1993) Tumor necrosis factor (TNF-α) downregulates the expression of endothelial nitric oxide synthase. Naunyn Schmiedebergs Arch Pharmacol 347 [Suppl]: R64

Giovanelli J, Campos KL, Kaufman S (1991) Tetrahydrobiopterin, a cofactor for rat cerebellar nitric oxide synthase, does not function as a reactant in the oxygenation of arginine. Proc Natl Acad Sci USA 88: 7091–7095

Heinzel B, John M, Klatt P, Böhme E, Mayer B (1992) Ca^{2+}/calmodulin-dependent formation of hydrogen peroxide by brain nitric oxide synthase. Biochem J 281: 627–630

Hevel JM, White KA, Marletta MA (1991) Purification of the inducible murine macrophage nitric oxide synthase. Identification as a flavoprotein. J Biol Chem 266: 22789–22791

Hibbs JB Jr, Taintor RR, Vavrin Z (1987) Macrophage cytotoxicity: role for L-arginine deiminase and imino nitrogen oxidation to nitrite. Science 235: 473

Hope BT, Michael GJ, Knigge KM, Vincent SR (1991) Neuronal NADPH diaphorase is a nitric oxide synthase. Proc Natl Acad Sci USA 88: 2811–2814

Janssens SP, Shimouchi A, Quertermous T, Bloch DB, Bloch KD (1992) Cloning and expression of a cDNA encoding human endothelium-derived relaxing factor/nitric oxide synthase. J Biol Chem 267: 14519–14522

Klatt P, Heinzel B, John M, Kastner M, Böhme E, Mayer B (1992) Ca^{2+}/calmodulin-dependent cytochrome c reductase activity of brain nitric oxide synthase. J Biol Chem 267: 11374–11378

Kwon NS, Nathan CF, Stuehr, DJ (1989) Reduced biopterin as a cofactor in the generation of nitrogen oxides by murine macrophages. J Biol Chem 264: 20496–20501

Lowenstein C, Glatt CS, Bredt DS, Snyder SH (1992) Cloned and expressed macrophage nitric oxide synthase contrasts with the brain enzyme. Proc Natl Acad Sci USA 89: 6711–6715

Lyons CR, Orloff GJ, Cunningham JM (1992) molecular cloning and functional expression of an inducible nitric oxide synthase from a murine macrophage cell line. J Biol Chem 267: 6370–6374

Marsden PA, Schappert KT, Chen HS, Flowers M, Sundell CL, Wilcox JN, Lamas S, Michel T (1992) Molecular cloning and characterization of human endothelial nitric oxide synthase. FEBS Lett 307: 287–293

Mayer B, Heinzel B, Klatt P, John M, Schmidt K, Böhme E (1992) Nitric oxide synthase-catalyzed activation of oxygen and reduction of cytochromes: reaction mechanisms and possible physiological implications. J Cardiovasc Pharmacol 20 [Suppl 12]: S54–S56

Michel T, Li GK, Busconi L (1993) Phosphorylation and subcellular translocation of endothelial nitric oxide synthase. Proc Natl Acad Sci USA 90: 6252–6256

Nakane M, Mitchell J, Förstermann U, Murad F (1991) Phosphorylation by calcium calmodulin-dependent protein kinase II and protein kinase C modulates the activity of nitric oxide synthase. Biochem Biophys Res Commun 180: 1396–1402

Nakane M, Schmidt HHHW, Pollock JS, Förstermann U, Murad F (1993) Cloned human brain nitric oxide synthase is highly expressed in skeletal muscle. FEBS Lett 316: 175–180

Nathan C (1992) Nitric oxides as a secretory product of mammalian cells. FASEB J 6: 3051–3064

Palmer RMJ, Ashton DS, Moncada S (1988) Vascular endothelial cells synthesize nitric oxide from L-arginine. Nature 333: 664–666

Pollock JS, Förstermann U, Mitchell JA, Warner TD, Schmidt HHHW, Nakane M, Murad F (1991) Purification and characterization of particulate endothelium-derived relaxing factor synthase from cultured and native bovine aortic endothelial cells. Proc Natl Acad Sci USA 88: 10480–10484

Pollock J, Klinghofer V, Förstermann U, Murad F (1992) Endothelial nitric oxide synthase is myristilated. FEBS Lett 309: 402–404

Prónai L, Ichimori K, Nozaki H, Nakazawa H, Okino H, Carmichael AJ, Arroyo CM (1991) Investigation of the existence and biological role of L-arginine/nitric oxide pathway in human platelets by spin trapping/EPR studies. Eur J Biochem 202: 923–930

Schmidt HHHW, Klein MM, Niroomand F, Böhme E (1988a) Is arginine a physiological precursor of endothelium-derived nitric oxide? Eur J Pharmacol 148: 293–295

Schmidt HHHW, Nau H, Wittfoht W, Gerlach J, Prescher K-E, Klein MM, Niroomand F, Böhme E (1988b) Arginine is a physiological precursor of endothelium-derived nitric oxide. Eur J Pharmacol 154: 213–216

Schmidt HHHW, Pollock JS, Nakane M, Gorsky LD, Förstermann U, Murad F (1991) Purification of a soluble isoform of guanylyl cyclase-activating-factor synthase. Proc Natl Acad Sci USA 88: 365–369

Schmidt HHHW, Pollock JSP, Nakane M, Förstermann U, Murad F (1992a) Ca^{2+}/calmodulin-regulated nitric oxide synthases. Cell Calcium 13: 427–434

Schmidt HHHW, Smith RM, Nakane M, Murad F (1992b) Ca^{2+}/Calmodulin-dependent NO synthase type I: a biopteroflavoprotein with Ca^{2+}/calmodulin-independent diaphorase and reductase activities. Biochemistry 31: 3243–3249

Schmidt HHHW, Warner TD, Ishii K, Sheng H, Murad F (1992c) Insulin-secretion from pancreatic B-cells caused by L-arginine-derived nitrogen oxides. Science 255: 721–723

Schmidt HHHW, Warner TD, Nakane M, Förstermann U, Murad F (1992d) Regulation and subcellular location of nitrogen oxide synthases in RAW264·7 macrophages. Mol Pharmacol 41: 615–624

Schmidt HHHW, Gagne GD, Nakane M ,Pollock JS, Miller MF, Murad F (1992e) Mapping of NO synthase in the rat suggests colocalization with NADPH diaphorase but not soluble guanyl cyclase and novel paraneural functions for nitrinergic signal transduction. J Histochem Cytochem 40: 1439–1456

Schmidt HHHW, Lohmann SL, Walter U (1993) The nitric oxide and cGMP signal-transduction pathway. Biochim Biophys Acta 1178: 153

Sessa WC, Harrison JK, Barber CM, Zeng D, Durieux ME, D'Angelo DD, Lynch KR, Peach MJ (1992) Molecular cloning and expression of a cDNA encoding endothelial cell nitric oxide synthase. J Biol Chem 267: 15274–15276

Snyder SH, Bredt DS, (1991) Nitric oxide as a neuronal messenger. Trends Pharmacol Sci 12: 125-128

Stuehr D, Griffith O (1992) Mammalian nitric oxide synthases. Adv Enzymol Related Areas Mol Biol 65: 287–346

Stuehr DJ, Cho HJ, Kwon NS, Nathan CF (1991) Purification and characterization of the cytokine-induced macrophage nitric oxide synthase: an FAD- and FMN- containing flavoprotein. Proc Natl Acad Sci USA 88: 7773–7777

Tayeh MA, Marletta MA (1989) Macrophage oxidation of L-arginine to nitric oxide, nitrite, and nitrate: tetrahydrobiopterine is required as a cofactor. J Biol Chem 264: 19654–19658

Wilcox CS, Welch WJ, Murad F, Gross SS, Taylor G, Levi R, Schmidt HHHW (1992) Nitric oxide synthase in macula densa regulates glomerular capillary pressure. Proc Natl Acad Sci USA 89: 11993–11997

Xie Q-W, Cho HJ, Calaycay J, Mumford RA, Swiderek KM, Lee TD, Ding A, Troso T, Nathan C (1992) Cloning and characterization of inducible nitric oxide synthase from mouse macrophages. Science 256: 225–228S

Subject Index

Current Topics in Microbiology and Immunology

Volumes published since 1989 (and still available)

Vol. 173: **Pfeffer, Klaus; Heeg, Klaus; Wagner, Hermann; Riethmüller, Gert (Eds.):** Function and Specificity of Á / ‰ TCells. 1991. 41 figs. XII, 296 pp. ISBN 3-540-53781-3

Vol. 174: **Fleischer, Bernhard; Sjögren, Hans Olov (Eds.):** Superantigens. 1991. 13 figs. IX, 137 pp. ISBN 3-540-54205-1

Vol. 175: **Aktories, Klaus (Ed.):** ADP-Ribosylating Toxins. 1992. 23 figs. IX, 148 pp. ISBN 3-540-54598-0

Vol. 176: **Holland, John J. (Ed.):** Genetic Diversity of RNA Viruses. 1992. 34 figs. IX, 226 pp. ISBN 3-540-54652-9

Vol. 177: **Müller-Sieburg, Christa; Torok-Storb, Beverly; Visser, Jan; Storb, Rainer (Eds.):** Hematopoietic Stem Cells. 1992. 18 figs. XIII, 143 pp. ISBN 3-540-54531-X

Vol. 178: **Parker, Charles J. (Ed.):** Membrane Defenses Against Attack by Complement and Perforins. 1992. 26 figs. VIII, 188 pp. ISBN 3-540-54653-7

Vol. 179: **Rouse, Barry T. (Ed.):** Herpes Simplex Virus. 1992. 9 figs. X, 180 pp. ISBN 3-540-55066-6

Vol. 180: **Sansonetti, P. J. (Ed.):** Pathogenesis of Shigellosis. 1992. 15 figs. X, 143 pp. ISBN 3-540-55058-5

Vol. 181: **Russell, Stephen W.; Gordon, Siamon (Eds.):** Macrophage Biology and Activation. 1992. 42 figs. IX, 299 pp. ISBN 3-540-55293-6

Vol. 182: **Potter, Michael; Melchers, Fritz (Eds.):** Mechanisms in B-Cell Neoplasia. 1992. 188 figs. XX, 499 pp. ISBN 3-540-55658-3

Vol. 183: **Dimmock, Nigel J.:** Neutralization of Animal Viruses. 1993. 10 figs. VII, 149 pp. ISBN 3-540-56030-0

Vol. 184: **Dunon, Dominique; Mackay, Charles R.; Imhof, Beat A. (Eds.):** Adhesion in Leukocyte Homing and Differentiation. 1993. 37 figs. IX, 260 pp. ISBN 3-540-56756-9

Vol. 185: **Ramig, Robert F. (Ed.):** Rotaviruses. 1994. 37 figs. X, 380 pp. ISBN 3-540-56761-5

Vol. 186: **zur Hausen, Harald (Ed.):** Human Pathogenic Papillomaviruses. 1994. 37 figs. XIII, 274 pp. ISBN 3-540-57193-0

Vol. 187: **Rupprecht, Charles E.; Dietzschold, Bernhard; Koprowski, Hilary (Eds.):** Lyssaviruses. 1994. 50 figs. IX, 352 pp. ISBN 3-540-57194-9

Vol. 188: **Letvin, Norman L.; Desrosiers, Ronald C. (Eds.):** Simian Immunodeficiency Virus. 1994. 37 figs. X, 240 pp. ISBN 3-540-57274-0

Vol. 189: **Oldstone, Michael B. A. (Ed.):** Cytotoxic T-Lymphocytes in Human Viral and Malaria Infections. 1994. 37 figs. IX, 210 pp. ISBN 3-540-57259-7

Vol. 190: **Koprowski, Hilary; Lipkin, W. Ian (Eds.):** Borna Disease. 1995. 33 figs. IX, 134 pp. ISBN 3-540-57388-7

Vol. 191: **ter Meulen, Volker; Billeter, Martin A. (Eds.):** Measles Virus. 1995. 23 figs. IX, 196 pp. ISBN 3-540-57389-5

Vol. 192: **Dangl, Jeffrey L. (Ed.):** Bacterial Pathogenesis of Plants and Animals. 1994. 41 figs. IX, 343 pp. ISBN 3-540-57391-7

Vol. 193: **Chen, Irvin S. Y.; Koprowski, Hilary; Srinivasan, Alagarsamy; Vogt, Peter K. (Eds.):** Transacting Functions of Human Retroviruses. 1995. 49 figs. Approx. IX, 240 pp. ISBN 3-540-57901-X

Vol. 194: **Potter, Michael; Melchers, Fritz (Eds.):** Mechanisms in B-cell Neoplasia. 1995. Approx. 152 figs. XXV, 458 pp. ISBN 3-540-58447-1

Vol. 195: **Montecucco, Cesare (Ed.):** Clostridial Neurotoxins. 1995. Approx. 28 figs. Approx. 260 pp. ISBN 3-540-58452-8

Springer-Verlag
and the Environment

We at Springer-Verlag firmly believe that an international science publisher has a special obligation to the environment, and our corporate policies consistently reflect this conviction.

We also expect our business partners – paper mills, printers, packaging manufacturers, etc. – to commit themselves to using environmentally friendly materials and production processes.

The paper in this book is made from low- or no-chlorine pulp and is acid free, in conformance with international standards for paper permanency.

Printing: Saladruck, Berlin
Binding: Buchbinderei Lüderitz & Bauer, Berlin